# Urban Emergency Management

# Urban Emergency Management

## Planning and Response for the 21st Century

**Thomas Henkey**

ELSEVIER

Butterworth-Heinemann
An imprint of Elsevier

Butterworth-Heinemann is an imprint of Elsevier
The Boulevard, Langford Lane, Kidlington, Oxford OX5 1GB, United Kingdom
50 Hampshire Street, 5th Floor, Cambridge, MA 02139, United States



**British Library Cataloguing-in-Publication Data**
A catalogue record for this book is available from the British Library

**Library of Congress Cataloging-in-Publication Data**
A catalog record for this book is available from the Library of Congress

ISBN: 978-0-12-805307-2



*Publisher*: Candice Janco
*Acquisition Editor*: Candice Janco
*Editorial Project Manager*: Hilary Carr
*Production Project Manager*: Vijayaraj Purushothaman
*Cover Designer*: Mark Roger

Typeset by MPS Limited, Chennai, India

# Contents

# About the Author

Thomas Henkey served for 6 years as the Senior Emergency Management Coordinator for the City of Chicago, where he was responsible for disaster planning and response, as well as special events, physical-security, infrastructure, transportation, and antiterrorism analysis. He also has nearly 15 years of experience in a range of private-sector and nonprofit safety and security management roles. He is currently the Director of Emergency Management for Titan Security Group.

He is a Certified Emergency Manager (CEM), a Certified Institutional Protection Manager (CIPM II), and a member of the International Association of Emergency Managers, the ASIS Cultural Properties Council, the International Foundation for Cultural Property Protection, the Illinois Security Professionals Association, and the Chicago Council on Global Affairs. He is the Vice Chairman of the Chicago Cultural Properties Security Group, and the former chairman of the ASIS International Museum Committee. He holds undergraduate degrees from St. Louis University and a master's degree in Emergency and Disaster Management from American Military University.

He lives in Chicago with his wife Megan, an educator and curriculum professional, and also enjoys decidedly nonemergency activities including music, literature, current events, motorcycles, hiking, and rockclimbing.

# Preface

This book began, as so many projects do, by happy accident. I was introduced to the Elsevier team when discussing an entirely different project with author and fellow safety and security professional Phil Purpura. Fortunately, after a couple of great conversations about emergency management and homeland security, Editor Sara Scott was gracious enough to review a few proposed project ideas. Together, we could not think of a single authoritative text focused upon emergency management in the urban environment.

And suddenly, I had a brand-new topic to research, draft, and edit—all in less than 2 years. In retrospect, I wouldn't have had it any other way. The process has been incredibly rewarding and educational, allowing me to both broaden and deepen my previous knowledge. But at the outset, it seemed as though I may have an impossibly steep hill to climb.

Soon, the text organically divided itself into three distinct and roughly equal portions. The first segment, consisting of Chapters 1–4, is a strategic assessment that sets the stage for the remainder of the book. It begins with an overview and history of urban emergency management, and proceeds to lay a foundation by reviewing the three major types or causations of disaster.

The second segment includes Chapters 5–8, and gets into the systems and methodologies that comprise the emergency manager's toolkit. These include crucial planning strategies, incident command systems, and cross-sector collaboration. This portion of the text takes a deeper dive into the planning, organization, and interconnectedness of the field.

The third major segment includes five chapters focused on the phases of emergency management—prevention, preparedness, mitigation, response, and recovery. This section not only examines the traits of each phase but also offers practical, real-world guidance for managing each. The book concludes with a forward-looking chapter focused on the future of the field.

In order to provide the reader with maximum content and context, the full text features keywords, charts, graphs, tables, photos, and several brief case studies. Each chapter concludes with a summary and a detailed reference bibliography to provide additional sources and resources for further study. In essence, the text starts at a strategic level and becomes more tactical as it unfolds.

Ultimately, I sought to strike a balance between academia and operations. As with most emerging fields, emergency management and disaster science lack adequate interaction between these two vital communities. The best possible result for this project would be to start a conversation that increases and strengthens the ties among and between academics and practitioners within the field.

Finally, I sought to strike a contextual balance which would benefit both emergency management practitioners and intermediate-to-advanced students of the field. In the primary text, I strove for a tone that was both confident and conversational; formal, but not too formal.

This project was indeed a challenging and time-consuming one. Yet it was also extremely edifying and rewarding. I sincerely hope that it will motivate readers to become even more interested in the emerging fields of emergency management, homeland security, and the disaster sciences. As rapidly evolving professions, each can benefit from both proven concepts and new ideas. If *Urban Emergency Management* starts just one conversation or collaboration that helps to save a single life, then this has been time extremely well spent.

# Acknowledgments

There are enough worthy acknowledgments for this project to make up a book all their own, but I will try to keep it brief. As mentioned in the preface, I owe a debt to author Phil Purpura and Elsevier Acquisitions Editor Sara Scott for the happy coincidence that helped distill the idea and get this concept off the ground. Project Editor Hilary Carr then got me across the finish line, granting me an amazing amount of creative control. I am eternally grateful for their guidance and trust, and look forward to our next opportunity to collaborate.

A number of professional organizations over the years have also played a huge role in my development as an operator and a leader. These include the world's largest security organization ASIS International as well as the ASIS Cultural Properties Council and the local Chicago chapter. I am thankful for the experts and professionals of the Chicago Council on Global Affairs, the Chicago Cultural Properties Security Group, the Chicago Hotel, Entertainment and Tourism (HEaT) security group, the Chicago Public-Private Task Force, the Illinois Security Professionals Association (ISPA), the International Association of Emergency Managers (IAEM), and the International Foundation for Cultural Property Protection (IFCPP). I am also very fortunate to have been associated with the ChicagoFIRST organization, and the Building Owners and Managers Association (BOMA) security and emergency preparedness committees for many years here in Chicago. Each and every one of these associations have served to grant me access to the finest professional minds and ideas in the field—a priceless privilege.

I gratefully acknowledge Professor Emeritus and all-around Renaissance Man Avis Meyer, who taught me how to recognize quality writing, among so many other valuable lessons. I thank writer and editor Richard Byrne and the *Riverfront Times* for taking a chance on me and providing my first paid writing gig years ago—as a music critic, no less. I am indebted to writer, teacher, and friend Lisa Hemminger for sharing her effusive love of words and language.

I have also been fortunate to have a litany of incredibly skilled, talented, and patient bosses over the years who have been willing to share their experiences with me. Within just the past decade, I must thank Jim Taff at Premier Security, as well as Mark Hubatch and Peter Moreau at AlliedBarton Security. I am also greatly indebted to Bill Schatz, Earl Zuelke, Jody Chattin, Rich Guidice, Bob Troy, Ray Orozco, Jose Santiago, Dan Ellis, and Gary Schenkel at the Chicago Office of Emergency Management and Communications (OEMC). At the Art Institute of Chicago, I learned from Michelle Lehrman, Russ Collett, and David Thurm. And as I take on new career challenges, I must express enormous gratitude to Dave Pack and Mark Hubatch (again) at Titan Security Group for entrusting me with our new venture.

And to my fellow managers and coordinators in the security and emergency management fields, please know that I was and am learning from you each and every day—even when I appeared to be too busy jumping from project to project to truly appreciate the hard work of my colleagues. The OEM team at the City of Chicago, and that city's nonprofit and private-sector security professionals, are some of the most amazing and skilled individuals I have ever encountered, and it has been an honor to work alongside them.

My immediate family has also been a great source of inspiration. My father had a career as a talented writer and editor, and my mother—like most of her family—as a skilled and capable educator. The apple truly does not fall far from the tree, and their inquisitive minds and voracious reading habits unquestionably contributed to mine, and to this text.

Finally, this book would not have been possible without one very special person and one very special group, and to them it is dedicated. The individual is my wife Megan, whose relentless curiosity and pursuit of both formal and informal knowledge—including her doctorate of education degree—has truly been an inspiration. And let us face it, this book is merely a blatant attempt to keep pace with her achievements.

The group dedication goes to all of the first responders who put their lives on the line with each and every shift and on each and every deployment. No matter what uniform you wear, no matter which jurisdiction or population or organization you protect, thank you for all that you do. May you always return safely to quarters and return safely to your friends and family.

CHAPTER 1

# Overview of Urban Emergency Management

## INTRODUCTION

On the evening of October 17, 1989, the weather was clear and seasonable in the San Francisco Bay area. A tangible excitement hung in the air as the two local Major League Baseball franchises—the San Francisco Giants and the Oakland A's—prepared to face one another in Game 3 of the World Series. Just after 5:00 p.m. local time, as Candlestick Park was still filling with fans, two tectonic plates shifted beneath a peak of the Santa Cruz Mountains known as Loma Prieta, a short distance away.

The significant quake measured 6.9 on the Richter scale, caused significant damage to local infrastructure, and resulted in 63 fatalities. Because of the intense media coverage of the World Series, the earthquake became the first to be covered live on US network television. It also became one of the first large-scale tests for the emerging field of expertise known as emergency management.

The Loma Prieta earthquake, as it came to be known, was far from the most destructive event in modern times. It caused neither the most fatalities nor the greatest financial damage of such events in the latter portion of the 20th century. However, due in large part to its location, timing, and the presence of the world's media, it represents an excellent point of reference in the development of urban crisis management.

The response and recovery phases of the earthquake took place under national and international scrutiny. In addition to the fatalities, there was damage to infrastructure, including utility, housing, and transportation assets. Difficult urban search-and-rescue operations took place at multiple locations, and the recovery process was both lengthy and costly. The damage to infrastructure helped motivate and accelerate legislation and retrofitting efforts related to seismic standards.



Urban Emergency Management. DOI: http://dx.doi.org/10.1016/B978-0-12-805307-2.00001-9

This natural disaster also helps to frame several key questions regarding emergency management in an urban setting. What comprises emergency management? How is the management of crises different in cities as compared to rural or wildland locations? What are the phases or stages of urban emergency management? And what qualities can emergency management offer to the overall goal of public safety?

## CONTEXT

### The Big Picture

As long as there have been human societies, there have been emergencies. And as long as there have been emergencies, there have been individuals who sought to manage them. Human society is essentially built upon the premise that we are stronger in the face of adversity as a group than we are as individuals. Early societies faced many of the same natural and weather-related hazards as we do today. Ancient historical and religious texts refer to plagues, floods, and earthquakes. Fortified cities emerged as protection against primarily human threats, in locations carefully chosen to counter hazards ranging from floods to marauders.

As states and nations took hold, they built increasingly dense and complex infrastructure as new technologies and materials became available. The Industrial Revolution added a new array of technological hazards to which cities had to react and adapt. In a very real sense, emergencies and crisis managers have been marching alongside civilization from the very beginning.

Modern emergency management began to truly emerge in the United States in the civil-defense efforts of the 1950s and 1960s. The threat of a catastrophic nuclear war between world superpowers necessitated an extensive organization to assure effective response and continuity of government. The military has continued as a key contributor to both the form and function of emergency management, with its hierarchal specialization and established chain of command providing a model for human-caused hazards in particular. Natural disasters, such as flooding and wildfires, soon drove the development of more effective crisis-management processes on multiple continents. (An extensive discussion of this period may be found in Chapter 2, Development and History of Emergency Management.)

Emergency managers have learned valuable lessons from related fields. Wildland firefighting can be defined as fire incidents that occur outside of significant human settlement, which have vegetation as their primary fuel source. The methods developed to battle such conflagrations also deserve extensive credit for developing the systems and mechanisms that form the core of modern emergency management. The forest fires endemic to the Western United States are complex and geographically large-scale incidents requiring extensive planning and coordination. The evolution of the National Incident Management System (NIMS) in the United States, and of other incident command systems the world-over, owe much to hard-fought lessons of these fire crews. (Such systems are reviewed in Chapter 6, Incident Command Systems and NIMS.) While wildfires are historically rural events, they have nonetheless informed the management of emergency incidents in the most urban of settings.

The evolution of emergency management can thus be seen as a long and ongoing process, including the adoption and adaptation of the best practices from multiple fields.

## The Urban Perspective

Translating lessons such as the Loma Prieta earthquake within the urban environment is still very much a work in progress. For this discussion, the term urban will refer to a geographic area with a comparatively high human population density and significant infrastructure. It is this density and complexity that pose significant challenges to emergency managers. Any large-scale or widespread damage is likely to affect multiple assets and systems.

Such areas may not encompass merely the traditional city boundaries, but also the suburbs in close proximity—including the slums or favelas that now surround city centers on multiple continents. In other words, as cities become denser, so will the terrain in immediate proximity.

By this definition, more than half the world's population already resides in urban areas. The World Health Organization (WHO) is among those tracking a rapidly urbanizing world. By 2014, WHO calculated that a full 54% of the Earth's human population lived in cities, and this population is expected to grow at nearly 2% annually through 2025. The presence of large numbers of people, and the systems necessary to support them, presents emergency managers with both unique opportunities and unique challenges.

As urban development pushes deeper into previously rural or wildland terrain across multiple continents, the lines defining what makes up urban, suburban, and rural landscapes continue to blur. A changing climate will only exacerbate such changes, bringing hazards across boundaries which were previously clear delineations.

To summarize, the key traits that may serve to differentiate urban emergency management from the more generalized field are:

- Higher population density

- More diverse population and linguistics
- Higher infrastructure density and complexity
- Higher property values per square mile/kilometer
- Higher likelihood of cascading impacts
- Greater number of responding agencies and organizations
- Higher concentration and shorter response time of media

Ultimately, an urban emergency manager is going to be dealing with a higher number of potential cascading impacts, and a shorter timeframe before such impacts degrade critical infrastructure or threaten the lives of civilians. Large-scale emergencies in public areas will quickly attract the attention of opinion-shaping stakeholders, including politicians and the media. Extensive attention will be given to these factors later in the text.

## Hazard, Threat, Risk, and Vulnerability

One of the most important differentiations in emergency management is that of hazards, risks, threats, and vulnerabilities. These factors may overlap and be the source of some confusion, but they nonetheless provide an important foundation for further discussion.

Hazard may be defined as a source or cause of potential harm or injury. Threat is an identified source of potential harm or injury, often implied to be manmade or intentional. Risk is the probability of harm or injury occurring. A vulnerability is an oversight or weakness which can be the method or avenue via which a hazard or threat causes actual damage or injury.

These concepts are certainly interrelated, and an example may help clarify the subtle differences. If a storm threatens a coastal town, wind and water are clear hazards. The failure of existing locks and levees may be additional threats. Vulnerabilities may include the susceptibility of infrastructure to damage, or a lack of redundant systems. Risk is the resulting probability of actual harm, injury, or death occurring when all factors are taken into consideration.

None of these factors exist in a vacuum, or independently of one another. In a traditional scientific sense, the concepts of hazard, threat, and vulnerability are all potential energy—possibilities waiting to happen. Risk is kinetic energy, the presumption of a given action or series of events being in motion.

While formulas vary somewhat, risk is often represented as the sum total of an asset's value combined with its threats and vulnerabilities. In other words, without any threats or vulnerabilities, there can be no risk. Chapter 5, Assessing Risk in Urban Areas, provides extensive information on this process.

### Natural, Human-Caused, and Hybrid Disasters

One of the most important differentiators within emergency management is the issue of core causation. As human beings, we like to put things into nice, neat buckets or categories.

The two major classifications commonly used to discuss and categorize disasters are natural and human-caused (sometimes referred to as manmade). As with any such broad categorization, they should be viewed as useful but limited generalizations.

A natural disaster is a crisis that is primarily caused by hazards that exist in, or are a result of, the natural environment. More specific types would include floods, earthquakes, wildfires, tornadoes, hurricanes, and tsunamis. The earthquake and resulting tsunami that struck the Indian Ocean on December 26, 2004 would be a powerful example of a natural disaster.

Human-caused or manmade disasters are increasingly commonplace. As the name implies, such crises are primarily the result of human intervention or error. Types would include technological and infrastructure failures, acts of terrorism, and structure fires. The aviation-borne terrorist attacks of September 11, 2001 would be an example of a human-caused disaster.

In recent decades, many disasters have defied easy categorization as purely natural or manmade in causation. Hybrid disasters are those that contain key elements of both natural and human elements, resulting in heightened risk or vulnerability. Types would include the impact of climate change on global weather patterns, or cascading events beginning in either the natural or manmade realm and spilling into the other. The earthquake, tsunami, and nuclear disaster that befell Fukushima, Japan in March of 2011 provides an example of a modern hybrid disaster.

Emergency managers in urban environment must be prepared for an expanding range of threats, and of hazard types. Further discussion and analysis of all three causative factor types may be found in Chapter 3, Natural Hazards and the Urban Environment, and Chapter 4, Human-Caused Hazards and Hybrid Hazards.

## PHASES OF EMERGENCY MANAGEMENT

A phased approach is one of the keys to effectively managing a crisis. While the onset of a given disaster can appear sudden, it is important to recognize that the highly visible response to an emergency situation is but one step in a continuous and ongoing process. Life-safety efforts will be underway before, during, and after a disaster.

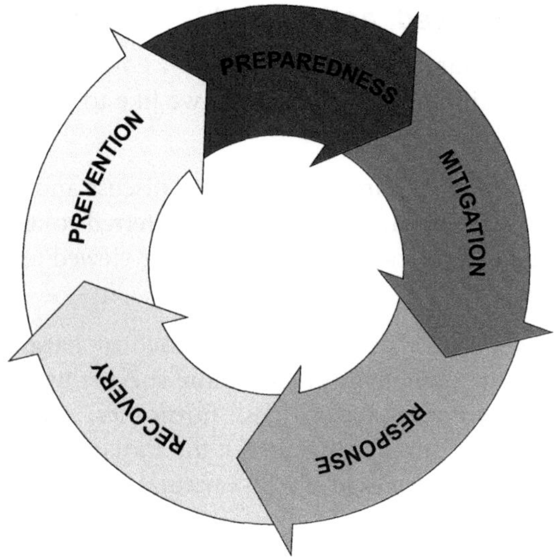


**FIGURE 1.1**
Modern emergency management now encompasses five major phases, including prevention, preparedness, mitigation, response, and recovery.

This emergency-management cycle now includes five phases or functions: prevention, preparedness, mitigation, response, and recovery (Fig. 1.1).

A key factor of this cycle is that it must be viewed as a continuum, with no defined beginning or ending point. It is also crucial to understand that phases may overlap or take place simultaneously. For example, prior to a disaster, preparedness and prevention efforts may be running in parallel. Immediately following a disaster, both response and recovery efforts may be ongoing.

Each phase will receive its own detailed analysis in a later chapter, yet a brief, introductory overview is worthwhile.

## Prevention

The most newly recognized phase of emergency management is prevention. While each of the other phases presumes that a disaster will or has taken place, the concept of prevention acknowledges that some crises can be completely avoided. The US Federal Emergency Management Agency (FEMA) recognizes prevention as one of its five core mission areas, defining it as encompassing the capabilities necessary to avoid, prevent, or stop a disaster from occurring. Such a capability would be largely focused on efforts to avoid technological or human-caused disasters, which are particularly common in urban areas.

## Preparedness

The emergency-management phase of preparedness—aligned with resilience and the protection mission area identified by FEMA—takes place largely prior to the onset of a disaster and includes training and planning for individuals and organizations. For this reason, it requires extensive participation by outside groups, and not merely emergency managers themselves. Such a capability applies to natural, human-caused, and hybrid disasters, and can have a significant effect on the impact or severity of a given hazard.

## Mitigation

Somewhat uniquely, the phase of mitigation has the consistent potential to occur simultaneously to all other phases of emergency management. Mitigation seeks to reduce the loss of life and property by lessening the overall impact of an incident or hazard. Effective mitigation efforts require effective hazard analysis and risk assessment. Such capacity has the potential to be highly relevant in terms of natural, human-caused, and hybrid hazards.

## Response

As the most visible phase of emergency management, response refers to actions taken to save lives, protect property and the environment, and meet basic human needs in the immediate aftermath of a disaster. Traditional law-enforcement, fire-suppression, emergency-medicine, and search-and-rescue functions all fall within this phase. Response is a universal reactive capability in cases of natural, human-caused, and hybrid crises.

## Recovery

Following a disaster, the phase of recovery intends to assist affected communities in effectively returning to some level of normal functionality. It is frequently one of the longest and costliest phases of emergency management, and is highly subject to public and political scrutiny. In a high-vulnerability environment, multiple recoveries may take place simultaneously. The recovery phase occurs following natural, human-caused, and hybrid disasters.

## CORE CONCEPTS OF EMERGENCY MANAGEMENT

### A Few Basics

With key terms and phases defined, it is possible to better understand the overarching function of emergency management. Managing an emergency or disaster may at first appear to be an oxymoron, as scholars including Sylves have noted previously (2008, p. 5). How do you control a crisis or catastrophe as it unfolds?

The short answer is of course that we really do not. The field of expertise is referred to as "management" rather than "command" or "control" for good reason. Yet much can be done prior to a disaster occurring, including efforts related to prevention, preparedness, and mitigation.

The true core of emergency management is compressing chaos. In the midst of a bona fide disaster, there will be a period of chaos and confusion during which the threat is not contained. Within this timeframe, the number of casualties and damage to infrastructure will outstrip the number and capabilities of first responders. Misinformation and rumors will swirl. And key organizational and political leaders will remain out of the loop, or worse still, make poor decisions based on bad information and rushed judgment.

An emergency manager's ultimate function is to reduce and limit this timeframe. We cannot control all aspects of a disaster, but we can squeeze the chaos into the smallest possible period of time.

One of the key tenants of emergency management is that a crisis or emergency is handled at the lowest possible authority level. In Western nations, this translates to responsibility being spread among individuals, insurance providers, private organizations, and various levels of government.

Drabek notes that the first response to any major event will be local (2007, p. 11). Effective emergency management programs emphasize horizontal or lateral connections in such circumstances, yet knowing how to rapidly and effectively request interorganizational, regional, state, tribal, or federal resources is absolutely key to the success of any emergency management network.

Emergency management requires a depth and breadth of expertise that may well be unequaled in other operational fields. Potential subject-matter expertise exhibited by emergency managers includes:

- Administration
- Civil engineering
- Communications
- Emergency medical services
- Finance

- Fire suppression
- Hazardous materials/chemistry
- Homeland security
- Hydrology
- Information technology
- Land-use planning and zoning
- Law enforcement
- Logistics/transportation
- Meteorology
- Mortuary services
- Psychology
- Public relations
- Public works
- Search and rescue
- Sociology
- Volunteer management

No one individual can be expected to possess true expertise across each and every one of these professional fields, but one individual may be required to efficiently coordinate and guide all of this collected knowledge. Such is the role of the emergency manager.

The position is generally seated within a governmental agency—be it local, regional, state, tribal, or federal. Many corporations and nonprofits now have an officer-level position with similar responsibilities within their own organizations. For example, the corporate roles of risk management and business continuity are closely related to emergency management. The academic fields of disaster science and disaster risk reduction also correlate extensively with emergency management.

In an operational context, the ideal and priority is to contain a crisis within the lowest level of command, and within the shortest amount of time. To visualize this concept, it is helpful to review a hypothetical example of a tanker truck that spills its load onto a public highway.

At its simplest, a well-trained driver or crew may be able to contain the spill using his or her own equipment and resources. A few small containers could merely be loaded back on the truck with a dolly. A small quantity of a non-toxic substance could be cleaned up with absorbent materials on the spot.

However, if the quantity is much larger, or the substance is highly toxic or flammable, the scenario changes significantly. A third-party hazardous-materials crew might be summoned, and the local fire department may be required to assist. If the substance ignites or spills into a nearby stream, further expertise and resources would be required. The highway may need to be closed, igniting nearby vegetation may become a risk, an extensive environmental cleanup may be required, or nearby residents may need to be temporarily evacuated to a safe distance. The impact to traffic and natural resources could be truly regional in scope.

This simple example displays the sliding scale that emergency managers face when planning for, and responding to, a crisis. The concept of vertical fragmentation also comes into play, as the possibility of ineffective or failed coordination between varying levels of government is a real possibility in emergencies spanning more than one level of oversight. If a given crisis spans more than one jurisdiction—a very common occurrence in the urban environment—the risk of horizontal fragmentation must also be considered. (Sadly, the television trope about two cops arguing about who has authority and jurisdiction over a case is not entirely myth.) In a large-scale disaster, an inefficiently coordinated governmental response may be both vertically and horizontally fragmented.

When assembling emergency plans for such hazards, the emergency manager must also be aware of multiple stakeholders. Stakeholders are individuals or groups with interests in a given project, program, or policy (Drabek, 2007, p. 13). Such interests may have extensive political or economic impacts of which the emergency manager must be cognizant. One of the most important tasks and skillsets for an emergency manager is being able to communicate complex concepts amongst multiple parties during highly stressful circumstances.

Along similar lines of thought, Sylves (2008, p. 21) lists core planning assumptions that decision makers must consider and encompass prior to a disaster taking place:

- Disaster planning is a continuous process and should be based on trends rather than a single emergency. Planning must allow for the constant incorporation of new findings.
- Disaster planning should attempt to reduce uncertainty in crises by anticipating problems and projecting potential solutions.
- Disaster planning stipulates that an appropriate response is more vital than a rapid response.
- Disaster planning is based on likelihood. Procedures need to address what people will actually do during emergencies, not on myths or nostrums about human behavior.
- Disaster planning involves education on response and recovery. People need to know that emergency procedures exist and that it is important to understand and follow these procedures.
- Disaster management needs to be "sold" effectively to communities to be taken seriously by the public.

- Disaster management requires exercises and practice, otherwise making the best plans worthless.

Once again, planning and preparation are absolutely crucial to effective emergency management. Any veteran emergency manager will attest that this is ultimately time well-spent, and directly correlates to the effectiveness of initial response and recovery efforts.

## Whole Community Planning

One of the most important emerging concepts in emergency planning is that of whole community. At its simplest, whole community is a means by which civilians, responders, officials, and organizational and community leaders can collectively strengthen their common assets, capacities, and interests (FEMA, 2011, p. 3). Ultimately, the concept is a cross-sector effort to plan for emergencies involving individuals with differing abilities. In the first decade of the 21st century, FEMA research, discussions, and outreach produced six whole community strategic themes:

- Understand community complexity
- Recognize community capabilities and needs
- Foster relationships with community leaders
- Build and maintain partnerships
- Empower local action
- Leverage and strengthen social infrastructure, networks, and assets

Ultimately, a given jurisdiction or organization is only as strong as its weakest link. Some residents, employees, and visitors will require additional assistance in case of emergency. Rather than reinventing the wheel, FEMA and its partners are attempting to leverage what works well in communities on a daily basis. (For more information on the whole community concept, see Chapter 10, Preparedness Phase.)

## Incident Command Systems

Handling a wide scope and scale of emergencies, and a wide range of potential stakeholders, requires an effective incident command system (ICS). ICS is one of the most comprehensive and effective tools at an emergency manager's disposal.

Responders have long recognized disaster as a chaotic and confusing series of related events. The chain of command was thus often tangled and unclear as well. In order to clarify and improve the situation, command models utilized by the military and wildland firefighting were again used as the basis for a new structure. At its core, ICS is a hierarchal, flexible, and modular methodology for command and control. Or as Erickson deftly states, "ICS therefore provides a flexible framework of management rather than a definitive algorithm" (2006, p. 50).

ICS divides authority for key functions among responsible subject-matter experts. In the United States, the structure is referred to as the National Incident Management System, or NIMS. The major categories utilized in its command structure are operations, finance and administration, planning, and logistics. For optimal efficiency, each unit maintains a strict span of control and ultimately reports to an overall incident commander. At its best, the system can also disentangle political and operational responsibilities. (NIMS and ICS are discussed in much greater depth in Chapter 6, Incident Command Systems and NIMS.)

## Emergency Management as a Profession

Readers of this text clearly have a vested interest in emergency management as students, practitioners, or both. What is the general definition of emergency management? And just what *is* a professional emergency manager?

Sylves provides an excellent definition. Emergency management is the discipline and profession of science, technology, planning, and management to address extreme events with the potential to injure or kill large numbers of people, do extensive property damage, and disrupt community life (2007, p. 5).

An emergency manager is thus the individual responsible for implementing and overseeing such a program. Traditionally, this has been a public-sector, governmental role. In rural areas, this is often a part-time role for someone with a related profession—a fire chief, sheriff, or even local judge. In urban environments, emergency managers are much more likely to be full-time professionals in a corporate, nonprofit, or governmental role.

Indeed, corporations and larger nonprofits have increasingly begun utilizing emergency managers within their organizational structures. As Erickson notes, proper emergency planning begins with the owners, managers, and operators of a given facility (2006, p. 67). For-profit companies and nonprofit organizations are recognizing that the concept of an emergency manager is not a foreign one, and may indeed directly benefit their operations. Such decisions are sometimes made at the behest of related entities including a board of directors or an insurance provider. In nongovernmental settings, the role may include phase-centric responsibilities such as risk management (preparedness and mitigation) or resilience and business continuity (recovery).

Further, emerging in tandem with the profession of emergency management is the academic field of study known as disaster science. This field includes elements of engineering, climatology, organizational management, sociology, economics, and communications. The interconnected field of disaster risk reduction similarly employs functions such as technology and data analysis. Just as a vast amount of depth and breadth of expertise is required to manage a complex incident, so too must that knowledge be methodically analyzed and expanded. This process may be as simple as generating an after-action report following a particularly challenging response, or as complex as a multiyear, multidisciplinary scientific study.

Additionally, particularly in the United States and Europe, this knowledge is becoming much more formalized in a rapid expansion of professional organizations and accredited degree programs. The International Association of Emergency Managers (IAEM) has a membership numbering in the thousands. Scores of colleges and universities, both traditional and online, now offer a wide range of undergraduate and graduate degrees in disaster science or emergency management.

In operational practice, these programs are no longer assembled ad hoc or "on the fly." Large American metropolitan areas such as New York, Chicago, Houston, and Los Angeles have dedicated emergency management departments (Fig. 1.2). Often referred to as an office of emergency management (OEM), these units operate with a large amount of autonomy and frequently report directly to the mayor or top elected official in a given jurisdiction. (Although such agencies are not completely immune from politics; in New York, the OEM has flip-flopped between reporting directly to the mayor or to the police commissioner.) Across the developed world, these agencies are also often responsible for answering and dispatching 000, 112, 911, or 999

**FIGURE 1.2**
The Chicago skyline with police and emergency management command vans in the foreground during a 2010 conference. *Credit: Author.*

emergency calls—depending what nation you are in—as well as emergency planning, antiterrorism analysis, and federal or regional grant disbursement.

Given the persistent—and some may argue increasing—global threat picture including terrorism and climate change, this trend is highly unlikely to be reversed anytime soon. While emergency management departments face the same corporate and governmental budget challenges as their fellow agencies, they have also proven their worth across a wide range of threats and hazards.

Whether on the practitioner or academic side of the profession—and whether engaged with the public, private, or nonprofit sectors—it is an undeniably dynamic and active time to be an emergency manager. The field continues to rapidly evolve, and is ripe for additional developments. (See Chapter 2, Development and History of Emergency Management, and Chapter 4, Human-Caused Hazards and Hybrid Hazards, for additional information.)

In essence, being an emergency manager may have slightly varied meanings within different organizations and different jurisdictions. Some managers may deal broadly with multiple risks and hazards, others with a single phase of the emergency-management cycle. Yet as the profession evolves, the consistent, core qualification remains technical expertise combined with a temperament capable of dealing effectively with extremely stressful incidents before, during, and after they occur.

## Management of All Hazards

The final core concept of emergency management to address in an initial overview is the all-hazards approach. Stakeholders expect emergency managers to be capable of handling virtually any crisis. The three actual, large-scale disasters referenced previously—the Indonesian tsunami, the September 11 terror attacks, and the Fukushima meltdown—have little in common. Many lives were lost, and many lives hung in the balance depending largely on the actions of first responders, emergency managers, and public-safety officials.

What these complex disasters did have in common was an expectation by victims and the whole community that help was on the way. That no matter what the original threat or hazard, there were experts and professionals ready and willing to expend every conceivable effort to compress the chaos.

This juncture is the point at which the field of expertise diverges somewhat from its military or wildfire roots. While military and firefighting units tend to be highly specialized—think combat engineer or bulldozer operator—emergency managers must also, on some level, be capable generalists.

Veteran emergency managers often come from related fields, including law enforcement, security, the fire services, or public health. And while they are expected to retain such expertise and apply it to their new role, they are also required to diversify and learn about the hazards, threats, and risks that are present within their own operation or jurisdiction.

The ultimate expectation of a professional emergency manager is that he or she possesses basic knowledge about a given hazard, and can unite true subject-matter experts in the effort. Ultimately, a skilled emergency manager is one of the highly trained responders running toward the metaphoric smoke and gunshots, with a well-designed operations plan ready to deploy. Such outcomes require extensive planning and coordination prior to a crisis, and are the ultimate basis for the courses of action presented throughout this text.

## SUMMARY

Emergency management is an established and rapidly evolving field. Learning important lessons from previous military, civil defense, and wildland firefighting efforts, the profession seeks to manage the chaos and confusion that accompany a major incident. It is vital to plan for a wide range of hazards and threats which can have both natural and manmade causes. Hazard, threat, vulnerability, and risk are all key factors which must be considered.

The rising percentage of the world's population living in urban centers serves to magnify disaster impacts. Ranges of stakeholders have an expectation that emergency management professionals are ready and prepared for any crisis which may affect their organization or jurisdiction. This goal is accomplished through a continuous cycle that includes prevention, preparedness, mitigation, response, and recovery.

## References

Drabek, T. (2007). *Emergency management and homeland security curricula: Contexts, cultures, and constraints*. Paper presented to Western Social Science Association. Alberta, Canada.

Erickson, P. A. (2006). *Emergency response planning for corporate and municipal managers*. Burlington, MA: Butterworth-Heinemann.

Federal Emergency Management Agency. (2011). *A whole community approach to emergency management: Principles, themes, and pathways for action*. Washington, DC: U.S. Federal Government.

Sylves, R. S. (2008). *Disaster policy & politics*. Washington, DC: CQ Press.

## Further Reading

World Health Organization. (2009). Available from: <http://www.who.int/gho/urban_health/situation_trends/en/>.

CHAPTER 2

# Development and History of Emergency Management

## INTRODUCTION

Disease. War. Famine. Terrorism. Volcanoes. Floods. The list goes on and on.

Humanity has seen it all over the millennia, and in many ways it is amazing our species has survived at all. In some form or fashion, we owe our collective existence to a group of people stretching back through each of those catastrophes; a group who somehow managed to guide themselves and their communities through one existential challenge after another.

They went by many titles. Chief and general. Prime minister and commander. President and director. But what they all had in common is that they were actively managing emergency situations. That proud tradition continues today in the form of professional emergency managers, who seek to safely guide their organizations and jurisdictions through times of trouble.

This chapter will seek to examine recent political and operational movements and trends, including multiple successes and failures, to reach a better understanding of the current state of the art of emergency management.

## ANALYSIS

### An Uneven Beginning

At its core, emergency management deals with risk. Predicting, modeling, mitigating, and—when possible—managing risk has been the constant of emergency management since inception. Early examples included both formal and informal agreements of mutual aid between adjacent farms, towns, and villages.

As society advanced through the Industrial Age, all response remained local, with factories banding together to train firefighters or towns funding maritime rescue crews, for example. The first real, coherent federal program of the type in the United States took place in 1803, with the national government



Urban Emergency Management. DOI: http://dx.doi.org/10.1016/B978-0-12-805307-2.00002-0

assisting in the rebuilding and recovery of a fire-ravaged town in New Hampshire (Haddow, Bullock, & Coppola, 2008, p. 2). Yet this and similar actions of the period appear to have been based on specific political appeals rather than on any broader policy initiative.

The organizations supporting emergency management have evolved greatly over the past two centuries. Early governmental emergency response within the United States continued as a largely improvised and informal affair, with the town fire department or police handling everything they could, and calling on neighboring communities only when necessary. However, a number of larger events—some might be called disasters—began to make these organizational structures grow in fits and starts. Even at a federal level, these response models evolved "largely on an ad hoc basis" (Hogue & Bea, 2006, p. 4–5).

The San Francisco earthquake of 1906 provides vivid examples from this period, as local, state, and federal politicians wrangled behind the scenes over authorities and funding even as parts of the young city still smoldered. No firm, systemic resolution was reached.

## Systems Emerge With the 20th Century

Just a couple of decades later, public works organizations such as the Tennessee Valley Authority began to reduce flooding during the Great Depression era, but largely as a secondary consequence to producing hydroelectric power and creating jobs. The Flood Control Act of 1934 did give the US Army Corps of Engineers a wide authority over flood-mitigation efforts, a responsibility that continues to this day.

The vastness of the Second World War led to equally large changes in technology and infrastructure as well as in military command and control structures. Many returning soldiers entered professions such as firefighting and law enforcement in their home nations, bringing a military mindset and organizational hierarchy with them.

The Federal Disaster Relief Act of 1950 represented an important leap forward. The US government then had a mechanism to provide federal resources such as equipment and supplies—as well as grant funding—to state and local governments dealing with a crisis. Crucially, it was the first nationwide legislation not limited to single type of threat or hazard. Soon after, in 1953, the Federal Civil Defense Administration (FCDA) was created to coordinate multiple agencies responding to large-scale natural disasters. The 1950s and 1960s saw at least four federal agencies with emergency responsibilities, including the Office of Emergency Planning. These were the clear forerunners of the Federal Emergency Management Agency (FEMA) that would later standardize multiagency response. Yet with many conflicting and overlapping departments and agencies, there would be no easy path to a US national model.

The 1960s were a decade fraught with multiple disasters in North America. A punishing coastal storm damaged hundreds of miles of Atlantic Ocean coastline in 1962. In 1964, a massive earthquake—at 9.2 on the Richter scale, it remains one of the largest ever—struck Alaska, generating an equally massive tsunami. Several large tropical storms also struck the southern United States, including Hurricane Betsy in 1965.

With the recurring theme of water damage and flooding stitching these events together, the federal government finally began establishing coherent national policies, foremost the National Flood Insurance Act of 1968. The resulting National Flood Insurance Program remains one of FEMA's cornerstone efforts to this day.

The Disaster Relief Act of 1974 followed, but it too addressed authorizations in a piecemeal fashion, leaving multiple federal agencies in charge of vaguely defined portions of disaster planning, response, and recovery. Nonetheless, it was precedent-setting legislation, establishing the emergency-declaration process, funding grant efforts for recovery, and emphasizing that multiple hazards should be addressed by federal agencies.

## Civil Defense Weighs In

During the same period, the Cold War grew more frigid and nuclear weapons became a real threat, making civil defense a true concern. Four primary focuses emerged: continuity of government; adequacy of critical resources and capacities; industrial mobilization for military response; and localized emergency protective and response measures in an attack (Hogue & Bea, 2006, p. 6). This combination of readiness for both natural and manmade crises was to become increasingly important in the years and decades to come.

This period also saw widespread construction of fallout shelters for political entities and federal agencies deemed vital to continuity of government. The fact that such shelters could also be utilized in cases of extreme weather appears purely secondary. Not a direct part of the national discussion today, civil defense was the homeland security of its day. As Sylves notes, minimizing the impacts of a nuclear exchange "became a principle focus of US disaster management" (2008, p. 48). 1950 saw the passage of two important pieces of legislation, the Civil Defense Act and the previously mentioned Federal Disaster Relief Act.

These acts did have the important effect of enabling a release of federal aid without a direct and specific authorization from Congress. And this relationship between civil preparedness and disaster management continued throughout the 1950s and 1960s, as the Cold War continued to drive national priorities for the US government. For example, the Federal Flood Insurance Act of 1956 was passed by Congress, but quietly died because it was never funded.

This so-called "dual-use approach" culminated in the Disaster Relief Act of 1966, which directly linked civil defense systems and mechanisms to threats from natural disasters. This period also saw the evolution of the Crisis Relocation Plan (CRP), which was a forerunner of the large-scale evacuation plans for the civilian population still in effect today.

In retrospect, the emergence of the FEMA within the national-response structure seems like a natural course of action. After all, having a single, common point of centralized command and information-sharing during a disaster situation would appear to be a highly desirable goal.

But the road to a more centralized and independent emergency-management function within the federal government was not a smooth or rapid one. In fact, as late as the mid-1970s, there was great debate at the national level as to whether the functions of five or more military and nonmilitary departments *could* even be combined.

## A Fundamental Shift in the United States

The 1970s saw two critical developments within the field. The Disaster Relief Act of 1974 helped reduce the gap between public and individual assistance following a disaster. It also codified the emergency-declaration process, which gave the President the ability to dispense federal assistance even when an event failed to meet the higher "disaster" standards.

As noted by Hogue and Bea, "President Jimmy Carter submitted to Congress, on June 19, Reorganization Plan No. 3 of 1978, which proposed the merger of five agencies from the Departments of Defense, Commerce, and Housing and Urban Development, as well as GSA, into one new independent agency, the Federal Emergency Management Agency" (2006, p. 13).

Thus, it was not until mid-1978 that the formation of FEMA even began to truly take shape. Not surprisingly, there were growing pains associated with the creation of such an organization, and near-constant assessment and evaluation was ongoing.

However, it is worth noting that the very creation of a federal emergency department with broad responsibilities and authorities was largely due to timing and a confluence of events. The Carter Administration, for all of its failings, was one that was dedicated to streamlining the federal government, and in returning significant decision-making authority to the Executive

Branch. The Mideast oil crisis and a revolution in Iran had already sparked foreign-policy fears. And then, in March of 1979, a commercial nuclear reactor at the Three Mile Island facility in Pennsylvania suffered a partial meltdown and created a public panic. The actual impact was comparably minimal due to successful containment, but the message to politicians was clear: this could have been really, really bad, and what are you doing about it?

And so it was in 1979 that President Jimmy Carter's administration combined the previous "alphabet soup" of federal agencies with disaster preparedness and response oversight into a single unified entity—the FEMA. This crucial initiative streamlined the federal process for both mitigating the effects of disasters resulting from a wide range of hazards, and for keeping the President and his administration informed of developments in the field via a direct chain of command. At the time of its creation, FEMA represented the single largest consolidation of preparedness efforts in the history of the United States (Sylves, 2008, p. 59).

The goals and objectives noted in those early documents were indeed transformative. The new agency had responsibility and authority for an incredibly wide group of hazards, ranging from severe weather and flooding to forest fires and nuclear fallout. The positioning of such agencies within the government structure is also important. Critically, the agency head reported directly to the Executive Branch of the federal government, and a new emphasis was placed on forecasting, preparing for, and mitigating future emergencies.

FEMA officially sprang into existence on March 31, 1979—less than a week after the incident at Three Mile Island. John Macy, formerly head of the Office of Personnel Management (OPM), became the new organization's first director (Haddow et al., 2008, p. 6). And yet even more changes were looming.

## A Modern Approach Emerges

As is often the case in American politics, the incoming administration of Ronald Reagan had a different set of priorities from its predecessor. During the 1980s, FEMA was largely tasked with functions including preparing the continental United States for a nuclear exchange with the Soviet Union, and continuity of government. Louis Guiffrida, a former military and counterterrorism man, was named the new FEMA director. His brusque style rubbed many the wrong way, and his reputation is mixed. It was a mixed-tempo period that saw a rise in environmental disasters but a relative dearth of severe weather events such as hurricanes. Guiffrida moved FEMA into a single unified headquarters and launched the multidisciplinary training site in

Emmitsburg, Maryland, which would evolve into the highly respected Emergency Management Institute (EMI).

At the end of Reagan's second term, the crucial Stafford Act of 1988—implemented in 1989—essentially replaced the previous civil defense regulation and guidance. At its core, the act authorizes the executive branch to issue major emergency and disaster declarations, sets broad eligibility criteria, and specifies the types of assistance available from the federal government (Sylves, 2008, p. 60). The range of aid defined by the Stafford Act includes assistance to state and local governments, debris removal, and grants to impacted individuals. Throughout this period however, FEMA remained highly dependent on contractors and volunteers to assist in the response and recovery phases of any major incident.

The Stafford Act has also been criticized for focusing mitigation efforts to take place only after a disaster has occurred, concurrent to the recovery period (Birkland, 2006, p. 112). This oversight was later corrected, to some degree, by the Disaster Mitigation Act of 2000.

In a very real sense, this period represents the struggle that has been at the core of emergency management since its very inception, which is the push and pull between a military-based manmade hazards model and a more civilian-oriented all-hazards model. Frequent changes of federal administrations and senior leaders only magnify such policy oscillations. While particularly prevalent in the United States, this same balancing act takes place within capitals across the developed world.

1989 proved to be another crucial year for American emergency management. In addition to the incoming administration of George H. W. Bush and the implementation of the Stafford Act, Hurricane Hugo struck the southern United States in September and the Loma Prieta earthquake rocked California less than a month later. FEMA was roundly criticized for its slow and anemic response in both instances. In fact, South Carolina's veteran Senator Ernest Hollings infamously referred to the agency as a "bunch of bureaucratic jackasses" (Haddow et al., 2008, p. 9). FEMA's very existence may well have been saved by the relatively low number of fatalities in each instance.

The 1992 Federal Response Plan represented another significant step forward in the overhaul process, but did not prevent FEMA from coming in for harsh criticism following the landfalls of Hurricane Andrew in Florida and Hurricane Iniki in Hawaii later that same year. In September of 1992, Congress mandated the National Academy of Public Administration (NAPA) perform a complete review of local, state, and federal response protocols. Released in 1993, the report provided a blunt assessment of capabilities and made a wide variety of recommendations regarding FEMA's leadership and operations. The changes that resulted were the most sweeping for the agency until the aftermath of the attacks on September 11, 2001.

And yet, by 1993, systemic change was indeed at hand. Through the 1990s, transformational FEMA administrator James Lee Witt placed an unprecedented emphasis on mitigation as a key part of national readiness. Witt, appointed by President Bill Clinton to run the troubled agency, was the first true emergency-management professional to hold the role. The scholar Roberts notes that "the agency underwent one of the most remarkable turnarounds in administrative history" (2006, p. 16). And the fundamental changes Witt instituted included forming three major phase-based "directorates" within the agency: mitigation, preparedness, and response/recovery.

The massive, multistate Midwest flood of 1993 immediately greeted Witt and tested his resolve. (It also gave the author his first field experience at filling and stacking sandbags.) Wildfires, another California earthquake, and a second round of flooding all followed within 2 years. And the FEMA began to truly prove its worth to both politicians and the American public with a series of competent and highly visible response, recovery, and mitigation measures.

Crucially, the agency also used federal mitigation and recovery funding to relocate people and structures away from floodplains, thus avoiding the previous trap of repeatedly rebuilding the same home or business in the same flood-prone area. Clinton rewarded Witt and FEMA by elevating the director role to a cabinet position—a technocratic but vital promotion.

The international terrorism that targeted the World Trade Center in 1993, the domestic terrorism that befell Oklahoma City in 1995, and a rise in similar events in Europe and the Middle East were a foreboding of what was to come. FEMA seemed to avoid assuming a lead role on the issue of terrorism during these years, yet did include it as a part of its larger all-hazards mission (Table 2.1).

## This Changes Everything

Of course, no summary of disaster management in the modern United States would be complete without examining the events of September 11, 2001, and their resulting creation, the Department of Homeland Security (DHS). The hijacking of four passenger airliners that morning would forever change the profession of emergency management. Within a few short hours, nearly 3000 people were dead from the largest act of terrorism in recorded history. Emergency managers from across the East Coast mobilized for the initial

**Table 2.1** Summary of Significant US Emergency-Management Legislation

| Year | Title | Summary |
|---|---|---|
| 1934 | Flood Control Act | Granted Corps of Engineers authority for flood control and mitigation |
| 1950 | Federal Disaster Relief Act | Allowed state and local governments to access federal resources |
| 1953 | Federal Civil Defense Administration | Created first federal coordinating agency |
| 1968 | National Flood Insurance Act | Created National Flood Insurance Program |
| 1974 | Disaster Relief Act | Began to form federal all-hazards model |
| 1978 | Reorganization Plan Number 3 | Evolved into the Federal Emergency Management Agency |
| 1988 | Stafford Act | Set disaster declaration procedure and federal grant/assistance process |
| 2002 | Homeland Security Act | Created the Department of Homeland Security |
| 2003 | Homeland Security Presidential Directive 5 | Instituted the National Incident Management System command structure |
| 2006 | Post-Katrina Emergency Management Reform Act | Recognized natural disasters and affirmed FEMA's multiphase responsibilities |

response alongside their counterparts from the fire services, law enforcement, emergency medicine, and the military.

The first wave of responders was soon followed by a national response and recovery effort, including highly skilled urban search-and-rescue teams, canine units, civil engineers, grief counselors, and volunteer coordinators. The days that followed were some of the nation's darkest, but proved the value and importance of a qualified range of emergency-management professionals as never before. Before the dust had even settled at Ground Zero in Manhattan, a massive policy change was underway within the US government.

DHS was officially created by the Homeland Security Act of 2002, under President George W. Bush, and former Pennsylvania governor Tom Ridge was named as its first secretary. The general organizational format of the new agency was adopted in November of that year, and by March 1 of 2003, more than 175,000 employees from 22 existing federal agencies were formally folded into DHS. FEMA Director Joe Albaugh announced his resignation the same day, and was replaced by political nominee Michael Brown.

As the largest government reorganization since the creation of the Department of Defense in the mid-1940s, DHS represented the amalgamation of more than 20 federal agencies into a single "super department." The struggle of incorporating such a broad range of agendas and potential hazards is also key. Meshing the sometimes very different worlds of national security and emergency management continues to be a prime area of contention—and scholarship—to this day.

What is not subject to debate is that the American emergency-management community immediately, and quite understandably, elevated terrorism among its list of potential hazards and threats. FEMA found itself as a part of the much larger homeland security apparatus, and large numbers of National Guard and Reserve troops were deployed overseas in new conflicts in Afghanistan and Iraq. Senior emergency-management positions within local and state agencies began to be filled by more law-enforcement professionals than those from the fire services, as had largely been the case previously. In a general sense, the pendulum had again swung from the wildland fire model back to the military model of pre-1979.

The American National Incident Management System (NIMS) variant of incident command systems (ICS) was adopted and rolled out nationally during these critical years. (See Chapter 6, Incident Command Systems and NIMS, for further details.) Several core features of the ICS model were highly desirable, including scalability, common operating procedures, and plain language.

Congressional funding for DHS and FEMA also skyrocketed during this period, resulting in a vast expansion of mission areas and core capabilities. As with any such growth spurt, there were instances of waste and fiscal impropriety. By the time that Michel Chertoff became DHS secretary in 2005 and began an assessment and overhaul process, FEMA's mission had shifted almost entirely to response and recovery.

Federal funding began leveling off and then declining in the second full decade of the 21st century. Both human-caused and natural hazards would have a profound impact on cities, even as the profession of emergency management moved through a generational shift.

## A UN Model

The United Nations (UN) does a wide variety of development work that may be considered as aligning with the emergency-management phases of preparedness or mitigation. This would include improving transportation networks between far-flung urban areas, or improving the availability of fresh water supplies to reduce the risk of water-borne disease. The UN also does extensive work with safely housing refugees during and following human conflicts and natural disasters across rural and urban areas.

In 1995, the Emergency Response Division (ERD) was created within the UN's larger Development Programme (UNDP) operation. Specialized teams designed for prompt field response to disasters were allocated fully 5% of the overall UNDP budget (Haddow et al., 2008, p. 257). Within a few years, the Disaster Reduction and Recovery Programme (DRRP) and Bureau of Crisis Prevention and Recovery (BCPR) were also stood up under the larger UNDP umbrella program.

In a matter of just a few years, the UN had effectively become a formal, and major, player in international mitigation, response, and recovery phases. Mitigation efforts include the aforementioned infrastructure improvements, in addition to funding and capacity increases for including risk reduction in current and future development projects.

UN field-response efforts are overseen by the Office of Coordination of Humanitarian Affairs (OCHA). Typically 90% of the response budget for a specific disaster is received via targeted donations from individuals or member states. Recovery efforts include early, skilled damage assessments following disasters and developing reintegration programs for displaced persons and refugees.

So while the United Nations "Blue Helmets" performing peacekeeping actions may be the first image to spring to mind about the organization, the UN does in fact provide a wide range of disaster services to the international community, including urban areas.

## A European Model

In 2001, the European Union launched its Civil Protection Mechanism, a cooperative agreement offering aid and assistance across traditional national borders. It soon established the Emergency Response Coordination Centre (ERCC), an emergency operations center with extensive monitoring and communications capabilities. Members of the compact include the 28 nations of the core EU, in addition to around a half dozen additional signatory nations. Unlike the American DHS, the European effort seems to tilt its efforts toward natural hazards. Flooding across central and southeastern Europe in 2013 and 2014 was met with a relatively effective response, although issues of coordination and timeliness certainly remain.

Interestingly, the agreement allows any nation to request assistance, including those outside of the continent. Recent deployments of assets have included crises in Haiti, Japan, and Nepal—all far removed from the European theater. The organization was also critical in coordinating the European logistical response to the African Ebola outbreak in 2014.

While the mechanism is certainly paying dividends for member nations and others that request assistance, Europe still lags behind the more unified federal efforts of the United States. Yet as the vast majority of crises on the continent are still handled by local or national teams, the agreement—and the ERCC in particular—provides effective monitoring, notification, and communications channels. Given the comparative youth of the current EU programs, progress has been substantial.

## CASE STUDY: HURRICANE KATRINA

### The Facts

The US National Hurricane Center became increasingly concerned as it tracked a tropical depression strengthening into a tropical storm during the week of August 22, 2005. On August 25, the storm had become a weak hurricane as it clipped the southern edge of Florida and made its way into the Gulf of Mexico. The National Hurricane Center and National Weather Service began issuing immediate warnings to local and state governments on the afternoon of August 26.

By August 27, fed by the warm waters of the Gulf of Mexico, Katrina had intensified to a large Category 3 hurricane churning northwestward toward the Louisiana and Mississippi coasts. Within hours, it had grown into a massive Category 5 storm with winds near 175 miles per hour. The weather agencies continued to issue warnings via phone, videoconference, and the media, providing a highly accurate track of where they anticipated the monster storm would make landfall. That track passed directly over the city of New Orleans—the only major American urban area located largely below sea level.

A system of levees and pumps reinforced New Orleans and the surrounding parishes, yet many of them dated to the 1930s. Many other areas along the Gulf Coast had even less protection. Yet ineffective leadership coordination and weak public messaging left many residents and business owners unsure whether to evacuate or to try and ride out the storm in place, as many had with previous storms. A mandatory evacuation order was issued by the mayor of New Orleans on the morning of August 28—barely 24 hours before the hurricane would reach the city.

Katrina, which had slightly weakened to a Category 3 storm, came ashore in southern Louisiana on August 29 along the west bank of the Mississippi River, mere miles from the city of New Orleans. The area sits approximately 3 feet above sea level; the storm surge of water pushed in front of Hurricane Katrina peaked at approximately 28 feet. Any structure or living thing near the shoreline was simply washed away.

In the first, tense hours after the storm came ashore, New Orleans residents and officials thought they had "dodged a bullet." Initial reports indicated minimal wind damage and some street flooding due to nearly 8 inches of rain, but the levees seemed to be holding and most of the pumps were still churning away. It was not to be.

The unique topography of New Orleans and the surrounding parishes is crisscrossed by the Mississippi River, Lake Pontchartrain, and multiple canals. One by one, urgent often panicked reports began to come into to local officials and the media of massive flooding in the streets. The levees and pumps were failing, one by one, and the entire region was being inundated by several feet of brackish storm waters. In the end, the storm surge was just too much for the stitched-together, largely obsolete system that supposedly protected the city. In all, more than 50 levees would be completely overtopped or fail outright. The city of New Orleans was drowning, and the coasts of Mississippi and Alabama would fare little better.

## Analysis and Investigation

Daylight on the morning of August 30 revealed the awful truth. Contrary to early reports, the Gulf Coast had taken a direct hit from a powerful hurricane with a massive storm surge. Large sections of the shoreline from Alabama to east Texas had simply been washed away by walls of water, and the urban center of New Orleans had been particularly hard hit.

Valiant search and rescue efforts by responders and civilians began immediately after the storm passed, yet these were hampered by the lack of a coherent and organized plan. Prestaged resources and supplies sat outside the disaster zone awaiting guidance via a confused chain of command. Improvisation quickly became the rule, as when a Coast Guard helicopter crew realized many residents were trapped in their attics by floodwaters, and flew inland to a Home Depot store to retrieve bigger axes in order to chop through rooftops.

Despite such dedicated efforts, the damage had been done. Katrina caused just over $100 billion in property damages, the most of any hurricane in history. Eventually, the storm would be blamed for approximately 1800 deaths—a decade later, various government agencies have still not agreed on a firm number. Media coverage of desperate survivors stranded at public facilities with no sign of government intervention or assistance flashed across the globe.

The list of failures that contributed to this result starts long before 2005. The US Army Corps of Engineers were ultimately responsible for maintaining the levee system in and around New Orleans. However, the agency was overseeing a patchwork of infrastructure in various states of disrepair dating back decades and was chronically underfunded.

The various levels of government also experienced severe vertical and horizontal fragmentation. There was poor and ineffective communication between local, state, and federal leaders prior to and during the crisis. While there were some personal connections that proved useful when the threat became clear, no level of government appears to have been confidently aware of what the layer above or below it was responsible for. Ultimately, this paralysis would directly contribute to the indecision about whether to evacuate civilians or attempt to shelter them in place.

On a national level, FEMA itself had shifted its focus largely to terrorism-related threats since 2001. While funding for natural hazards remained essentially flat, promotions for top talent and a major budget increase decisively favored operational units focused on human-caused hazards.

FEMA leadership was also woefully lacking. Director Michael Brown was a political appointee with no emergency-management experience or qualifications. A financial donor to the Bush election campaign, his prior role was as the leader of the International Arabian Horse Association. This weakness directly contributed to the delayed and ineffective launch of the federal operational response.

It must also be noted that public perception played a role in the scope and scale of the disaster. Recent hurricanes had been relatively tame events, with few if any casualties. Roberts notes that "during the twentieth century the level of assistance expected from the federal government before and after a disaster ratcheted upwards" (2006, p. 25). Everyone seemed to believe that the feds were going to ride in at the last possible second with amazing plans and resources—except for the feds.

And thus it was really complacency at all levels that sealed the fate of those who lost their lives in late August of 2005. Previous hurricanes had been manageable, the thinking went, so this one will too. Everything was going to work out alright, until it did not.

## Case Study Conclusions

State and local governments must accept a greater role in mitigation and initial response to disaster, including vertical and horizontal planning and interaction. FEMA must also continue to learn from Katrina. Roberts makes excellent recommendations for a three-pronged approach: focus tasks on emergency management, revive the all-hazards and all-phases process, and give local and state entities more responsibility for preparedness and response (2006, p. 28).

Citizens, businesses, and nonprofits must also be more involved in the planning and preparedness process. Katrina showed that procrastinating until the last moment to take action could be a fatal mistake, particularly in an urban setting. A more informed, capable, and prepared populace is every emergency manager's best ally.

Ultimately, the Gulf Coast and its residents had been betrayed by ineffectual and indecisive leadership. Great blame has correctly been placed upon the George W. Bush administration, yet the infrastructure of levee systems and pumps had been neglected for decades. In this sense, Hurricane Katrina may be considered a classic, if costly, lesson in hybrid disasters. The core cause was a natural occurrence, but the horrible, cascading impacts were greatly magnified by human error and complacency at multiple levels.

## THE STATE OF THE ART

In retrospect, 2005 certainly marked a low point in the history of emergency management. So where does the field of expertise stand today?

Katrina did cause significant changes within the US emergency-management community. Michael Brown quit FEMA, under pressure, shortly after the storm. David Paulison, a fire-services professional, was moved into the role. A qualified planner and responder, Paulison nonetheless inherited FEMA's horrid reputation at the time, and carried the baggage of the inherent suspicion and mistrust between the wildland fire and emergency-management camps. That said, he shepherded the agency through an exceedingly difficult time.

Multiple examinations and investigations into the ineffectual Katrina response were launched by various governmental bodies, and Paulison spent a lot of time sitting in front of Congressional committees and subcommittees. No fewer than six pieces of legislation resulted, including—most importantly—the Post-Katrina Emergency Management Reform Act of 2006. This act reconsolidates all phases of federal emergency management within FEMA and grants the agency considerable autonomy over its assets and contractors (Haddow et al., 2008, p. 24). Crucially, it also establishes a number of FEMA regional offices across the United States, providing key, nearby points of contact for state and local partners.

Meanwhile, state and local emergency managers all along the Gulf Coast were tasked with an enormous recovery and rebuilding operation. Critically, largely due to these experts on the ground, massive mitigation efforts sprung up alongside the rebuilding. This included what amounted to the complete reconstruction of the New Orleans infrastructure of pumps and levees to modern standards. The stated goal was to protect the urban center and its residents from a future Katrina-like Category 3 hurricane.

European and Asian efforts continued along similar lines, displaying a mix of proactive and reactive policies for local and regional emergency management. It is very much worth remembering that the earthquake and resulting tsunami that stuck the Indian Ocean late in 2004—not Katrina or the September 11 attacks—remains the largest and most catastrophic incident in modern times. The initial response was again lagging, due in part to the remoteness of many of the locations impacted by the immense waves. Yet the world's ultimate response was unprecedented in both scale and in the number of nations taking part.

In May of 2009, newly elected US President Barack Obama appointed Craig Fugate to lead FEMA. Fugate had been the emergency-management director for the state of Florida and had successfully guided local and state-level responses to a number of natural disasters, including several hurricanes. He pushed for the adoption of technology within FEMA and dealt with a number of natural and technological disasters during his multiyear tenure. Fugate is widely respected in the emergency-management community, and the planning for and responses to Hurricane Sandy in 2012 and Hurricane Matthew in 2016 received generally high marks. His willingness to remain in the high-pressure role for the entirety of both 4-year terms also brought a rare period of stability to a post that often turned over multiple times during a single administration. The qualifications and status of incoming senior federal emergency managers during a more-recent wave of nationalistic and populist governments across the West is as yet unproven.

Overall, the state of the art could thus be accurately described as "active." Hazards ranging from earthquakes and terrorism to flooding and bridge collapses continue to face the global community, just as they have for decades. Emergency management and its evolving ICS methodologies will play a key role in facing these challenges today and in the future.

## SUMMARY

And so, nearly two decades into a new millennium, the profession of emergency management remains a dynamic one, hopefully learning from past disasters and applying best practices to ones that will certainly occur in our lifetimes. The strength and structure of Western emergency-management programs and processes has changed over time and has taken vital cues from both the armed forces and wildland firefighting. In the United States, key leaders and pieces of legislation helped form the emergency-management function at the local, state, and federal levels. The terrorist attacks of September 11 resulted in the creation of the DHS and a shift toward terrorism and other human-caused hazards, while the Hurricane Katrina disaster helped shift the national posture back toward an all-hazards model.

## References

Birkland, T. A. (2006). *Lessons of disaster: Policy change after catastrophic events*. Washington, D.C.: Georgetown University Press.

Haddow, G., Bullock, J., & Coppola, D. (2008). *Introduction to emergency management* (Third edition.). Burlington, MA: Elsevier.

Hogue, H., & Bea, K. (2006). *Federal emergency management and homeland security organization: Historical developments and legislative options*. Washington, D.C: Congressional Research Service.

Roberts, P. (2006). FEMA after Katrina. *Journal of Policy Review*, June/July, 15–33.

Sylves, R. (2008). *Disaster policy & politics: Emergency management and homeland security*. Washington, D.C: CQ Press.

## Further Reading

U.S. Department of Homeland Security (2011). *National preparedness system*. Washington, D.C: Federal Emergency Management Agency.

CHAPTER 3

# Natural Hazards and the Urban Environment

## INTRODUCTION

On December 26, 2004, nearly 20 miles beneath the surface of the Indian Ocean, an 800 mile section of tectonic plate suddenly shifted. These massive ruptures are referred to by geologists as megathrusts, and they are known for producing the most violent earthquakes on the planet.

The Sumatra–Andaman earthquake, as it would become known, was just such a case. It was the third most-intense earthquake in modern history and was felt thousands of miles from its point of origin off the coast of northern Indonesia. The worst was yet to come.

The entire 800-mile underwater rift pushed one side of the seabed rapidly upward, displacing an enormous amount of liquid. One of the largest earthquakes in modern times had just created one of the largest tsunamis in modern times. The massive bulge of water pulsed out along the full length of the shift in the tectonic plates, originating not from a single point, but from a long line running roughly north to south.

Near the surface of the Indian Ocean, this mass of water would appear only as a small wave less than a meter high. However, this liquid pulse would be moving at incredible speeds—in excess of 400 miles per hour—until it approached landfall and shallower water. There, it slowed down but translated its force upward, creating successive walls of water up to 80 feet tall as they reached the shoreline.

The destruction was catastrophic, wiping entire towns and villages from the face of the Earth in Indonesia, Sri Lanka, and Thailand. Fatalities were confirmed as far away as the mainland of eastern Africa. Measurable surges in the tides were observed on opposite sides of the planet, including South Africa, Mexico, and Alaska.



Urban Emergency Management. DOI: http://dx.doi.org/10.1016/B978-0-12-805307-2.00003-2

In total, the tsunami is believed to have killed approximately 230,000 human beings, primarily in Indonesia, although many bodies were simply washed out to sea and a final count may never be reached. This grim statistic makes it the worst natural disaster since at least 1976. The United Nations estimated that the cleanup and recovery process would be the most costly in its history, and nations committed $14 billion dollars to assist. Private donations to nonprofit and charity groups likely exceeded even this amount. In a very real sense, the disaster recovery is still ongoing more than a decade later.

## ANALYSIS

The Indian Ocean Tsunami of 2004 may provide an extreme example of a natural disaster, but it is highly indicative of the force contained within our environment. Natural disaster is rivaled only by war and genocide as the greatest cause of premature death across human history. And the United Nations Office for Disaster Risk Reduction estimates that natural disasters cause losses of $250–$300 billion per year (Miles, 2015, p. 2).

### Risk Factors

Clearly, natural hazards are worthy of our complete attention. Before addressing specific types, common risk factors must be addressed. Risk factors might be seen as magnifiers or multipliers, exaggerating positive or negative impacts of a given hazard for certain groups.

For example, fatalities from Hurricane Katrina tended to be poor, disabled, and/or from minority populations. The natural hazards—a hurricane and storm surge—were the root cause of these fatalities. However, risk factors contributed greatly to such a high number of fatalities. The culture and social custom in many impacted communities, particularly among older or disabled residents, was to "ride out" the storm in place, as was done with previous hurricanes. And the extremely late issuance of a mandatory evacuation order for New Orleans left poor individuals without access to a personal car with no real transportation options.

Such social, cultural, and economic traits are referred to as risk factors. It is important to note that their influence can either increase or lessen the impact of a particular hazard or threat. A wealthy individual is more likely to have a personal vehicle and access to savings or credit, creating a multifaceted support mechanism not accessible to an individual with a lower income. In dense urban environments, these advantages or disadvantages also tend to be quite geographic in nature, dividing neighborhoods along economic lines. Similar groupings are commonplace for sociological or cultural identifiers.

Recognized risk factors include:

- Age
- Gender
- Race
- Religion
- Health
- Security
- Preparedness
- Literacy
- Wealth
- Credit
- Savings
- Insurance

As noted by Haddow, Bullock, and Coppola, poor or disabled individuals are more likely to be living on the brink of fiscal disaster on a daily basis and will exhaust their resources much more quickly than the wealthy or able-bodied (2008, p. 71). Once again, such differences will become all the more stark in an urban setting.

In essence, the common trait seen across all of these risk factors is whether an individual or household has a support network in place. Whether social, cultural, or economic in nature, having a safety net in position prior to an emergency or disaster is the most reliable predictor of how an individual or household will fare in the midst of a crisis as compared to their peers.

## TYPES OF NATURAL HAZARDS

Again, a natural hazard may be defined as a source or cause of potential loss with nature as its primary driver. The natural world is a complex system, and the types of hazards it is capable of generating varies widely.

In 2016, a study supported by the United Nations International Strategy for Disaster Reduction (UNISDR), natural hazards were determined to have been the primary cause for 62% of all disasters, and 80% of all related fatalities, in the prior 50-year period. Weather-related causes were the top cause of fatalities, at 45% of disaster-related deaths, with geophysical causes such as earthquakes and tsunamis resulting in 30% of recorded fatalities (2016, p. 2). The significant or more frequently occurring types are certainly worthy of further review.

### Earthquakes

An earthquake is a seismic event, a sudden shifting of rock below the surface of the earth. Such shifts can be brief and violent, or gradual and "rolling" in nature. They can occur under land masses, or beneath oceans. And they are among the most destructive of all natural hazards.

Geological science has identified a number of large masses, or plates, that make up the outer crust or shell of the planet. These plates have been established since the earth cooled billions of years ago, yet they have been shifting and sliding past, along, and over one another ever since. The planes upon which these large slabs meet are referred to as faults and are the source location of most earthquakes. Multiple smaller subfaults crisscross existing continents and snake across the sea floor. What we humans consider solid ground is, in other words, merely a series of big slabs or rock sitting atop a molten core.

The severity of a tremor event is frequently judged by two scales of measure, the Richter Scale and the Mercalli Scale. Both have been updated and modified over time and measure slightly different factors. The Richter Scale is generally recorded by a seismograph and determines the amount or severity of shaking at a point on the earth's surface. It ranges from 1 (barely detectable) to 10 (cataclysmic). The Mercalli Scale is based on observational damage at the surface, ranging from I (no damage) to XII (catastrophic) (Table 3.1).

**Table 3.1** Comparison of Modern Mercalli and Richter Earthquake Scales

| Mercalli Scale | Descriptor | Damage Extent Narrative | Richter Scale |
|---|---|---|---|
| I–IV | Instrumental | No damage. | 4.3 or less |
| V | Rather strong | Damage negligible. Small, unstable objects displaced or upset; some dishes and glass broken. | 4.4–4.8 |
| VI | Strong | Damage slight. Windows, dishes, glassware broken. Furniture moved or overturned. Weak plaster and masonry cracked. | 4.9–5.4 |
| VII | Very Strong | Damage slight to moderate in well-built structures; considerable in poorly built structures. Furniture and weak chimneys broken. Masonry damaged. Loose bricks, tiles, plaster, and stones fall. | 5.5–6.1 |
| VIII | Destructive | Structural damage considerable, particularly to poorly built structures. Chimneys, monuments, towers, elevated tanks may fail. Frame houses moved. Trees damaged. Cracks in wet ground and steep slopes. | 6.2–6.5 |
| IX | Ruinous | Structural damage severe; some buildings will collapse. General damage to foundations. Serious damage to reservoirs. Underground pipes broken. Conspicuous cracks in ground; liquification. | 6.6–6.9 |
| X | Disastrous | Most masonry and frame structures/foundations destroyed. Some well-built wooden structures and brides destroyed. Serious damage to dams, dikes, embankments. Sand and mud shifting on beaches and flat land. | 7.0–7.3 |
| XI | Very disastrous | Few or no masonry structures remain standing. Bridges destroyed. Broad fissures in ground. Underground pipes completely out of service. Widespread earth slumps and landslides. | 7.4–8.1 |
| XII | Catastrophic | Damage nearly total. Large rock masses displaced. Lines of sight and level distorted. | 8.2 or more |

*Source: FEMA.*

Earthquakes pose a serious risk to infrastructure both above and below the surface and can collapse large or poorly constructed buildings and bridges. These cascading impacts can in turn sever utilities to large areas, or cause raging fires. And as noted in the Introduction, an underwater earthquake can create a tsunami, thus compounding and expanding the resulting scope and scale of the damage.

While earthquakes are somewhat predictable geographically—generally, but not always, occurring along identified fault lines—it is the timing and the cascading nature of the risk that proves extremely challenging. For example, flooding caused by a weakened levee may be difficult to contain due to simultaneous power outages. Fires caused by ruptured gas lines may be difficult to reach due to buckled streets. Thus it is not merely the intense shaking which must be addressed in effective urban emergency planning, but also the resulting impact on critical infrastructure.

Human activity can also contribute to seismic instability. Various academic researchers and the United States Geological Survey (USGS) examined a 300-fold increase in measurable earthquakes in the state of Oklahoma in a single decade and concluded that hydraulic fracturing for petroleum extraction (or "fracking") was a strong causative factor. This trend included a significant tremblor exceeding 5.0 on the Richter Scale in 2016. While natural processes remain a primary cause of earthquakes, further research is ongoing into potential human contributions.

Mitigation and preparedness are key portions of the emergency-management cycle for earthquakes. Leaders such as Japan have proven that strong building codes, including the reinforcement of older structures and preparedness drills for civilians and responders, are proven methodologies for reducing the damage and loss of life that result when an earthquake takes place.

## Floods

Floods are an excessive presence of water at ground level and are one of the single most destructive and costly natural hazards in human history. The two major categories of flooding are overland and riverine.

Overland flooding generally occurs when a very large amount of water inundates a geographical area in a short period of time. This could be due to an exorbitant amount of precipitation over a short period of time, for example, or due to a levee breaking and releasing a high volume of water.

Riverine flooding occurs near established bodies of water. It is generally slower developing than overland flooding, but can be even more catastrophic. Riverine flooding occurs when an extremely high volume of water enters an established body of water such as river, stream, or lake and exceeds it capacity.

It is worth noting that over millennia human development has clustered near navigable waterways, whether ocean harbors or inland riverways. Because such locations are so critical to providing a source of fresh food and water for a large population, and for commerce, most large urban areas are now located near large bodies of water.

So while natural forces remain the primary causative factor for flooding, humans again are a prime contributor. Extensive construction using impermeable materials such as concrete and blacktop prevent water from being absorbed into the ground. Levees and dams restrict or limit waterflow to existing channels, limiting absorption and evaporation and passing higher volumes to communities downstream. And the reduction of wetlands—nature's traditional flood-absorbing sponges—only serves to exacerbate the problem.

Also of note is the close relationship between flooding and other water-based natural disasters such as hurricanes and storm surges. Water is often the common factor during such disasters and becomes a cascading risk as it interfaces with the complex urban environment.

Flooding is frequently a long-term disaster. Its effects are felt long after the flood recedes, with large quantities of mud and debris frequently left in its wake, and the risk of water-borne disease increasing accordingly. Costs are also large and lingering. FEMA estimated specifically water-related expenses for the Great Midwest Flood of 1993 to be over $6 billion and for Hurricane Katrina in 2005 to be over $15 billion. Keep in mind that these numbers represent merely the flood-related costs and not tangential expenses.

Mitigation measures, such as strengthening levees and mandating effective storm drains for new construction, remain an urban emergency manager's best tool for combatting flooding. Planning for effective evacuation and sheltering mechanisms is also crucial for cities vulnerable to water-based threats.

## Hurricanes

Hurricanes are the fiercest type of tropical storms, with massive rotational walls of wind and water reaching 74 miles per hour or greater. Such storms form over open ocean waters and are referred to as typhoons in the Pacific Ocean and cyclones in the Indian Ocean. They are consistently one of the largest threats in terms of the size of the impacted geographical area.

Given their spawning over oceans, coastal cities are clearly at greatest risk from hurricane damage. Yet these storms are so large, they can travel hundreds of miles inland continuing to dump several inches of heavy rainfall as they slowly dissipate.

Hurricanes generally require warm water temperatures to gather strength. Thus hurricane season is from June 1 to November 30 in the Northern Hemisphere and November 1 to April 30 in the Southern Hemisphere. (Note that at tropical latitudes, cyclones have the potential to form all year long.) Once a tropical storm becomes a hurricane, it can last up to 2 weeks over open ocean waters.

Hurricane/cyclone/typhoon strength is gradated via the Saffir-Simpson scale, with higher numbers indicating higher wind speeds, and thus a more severe impact:

- Category 1—Wind speed of 74–95 miles per hour, with base storm surge of 4–5 feet
- Category 2—Wind speed of 96–110 miles per hour, with base storm surge of 6–8 feet
- Category 3—Wind speed of 111–130 miles per hour, with base storm surge of 9–12 feet
- Category 4—Wind speed of 131–155 miles per hour, with base storm surge of 13–18 feet
- Category 5—Wind speed of 155 + miles per hour, with base storm surge of 18 + feet

It is this combination of wind and water that makes all tropical storms, and hurricanes in particular, so dangerous. People, buildings, and infrastructure will all be gravely tested by such a combination of factors. Climate change is likely to increase both the frequency and intensity of such tropical storms.

Predictably, the cost and expense of damage resulting from these storms can be astronomical. Hurricane Katrina in 2005 remains the most costly disaster in American history, with total damage estimates of over $80 billion. (See the case study in Chapter 2, Development and History of Emergency Management.) This price tag may have been rivaled by Hurricane Sandy, had it not weakened substantially as it approached the New York and New Jersey coasts less than a decade later.

Typhoons in the western Pacific region reach even greater strength, with 1979's Typhoon Tip believed to be the largest and strongest storm in modern history based on atmospheric pressure, packing sustained winds of 190 miles per hour before slamming into southern Japan. In 2013, Typhoon Haiyan reached nearly as low a pressure reading, while reaching sustained winds of 195 miles per hour. Haiyan made landfall in the Philippines, causing at least 6300 fatalities.

Mitigation and preparedness measures again offer urban emergency managers the best defense against such massive storms. Computer hurricane-prediction models are becoming ever more accurate and useful, providing additional warning times for those in danger. There is no way to slow down or stop a force of nature with the strength of a tropical storm, one can only get out of the way or head to higher ground. Effective mass-notification systems and robust evacuation and relocation planning are absolutely crucial.

## Pandemics

A pandemic refers to a geographically widespread and uncontained outbreak of a communicable disease. History has recorded many such occurrences, and two of the worst were centered in Europe.

Approximately 75 million people perished during a bubonic plague outbreak that peaked in the later 14th century and was widely referred to as the "Black Death." Nearly 50 million were believed to have died during a 2-year influenza outbreak beginning in 1918—far more than in the preceding and better known First World War. Similar, although less catastrophic, outbreaks of each disease have occurred since.

Other existing diseases capable of such mass cross-infection include cholera, HIV/AIDS, malaria, and tuberculosis. Seasonal outbreaks of influenza are normally excluded from such terminology, but recent mutations such as the H5N1 strain, the so-called avian or bird flu, have proven to be far more virulent than most.

And though they have been known to science for decades, hemorrhagic fevers have recently received worldwide attention due to the 2013/2014 Ebola outbreak in western Africa. Such diseases are extremely transmissible and extremely deadly, often killing 90% of those who are infected that do not receive advanced, prompt medical care.

From an emergency-management perspective, pandemics represent a low-probability, high-risk threat. Communicable disease is much easier to spread from person to person in the urban environment, and via modern continent-hopping transportation systems. Initial alerts will likely be raised by local health providers and hospitals, or by federal health officials, such as the Centers of Disease Control and Prevention (CDC) in the United States.

Public-health officials will take the lead on any such pandemic event in the Western world, but emergency managers will play a crucial supporting role. Training and education for all public-safety officials and responders will be key to allow quick recognition of pandemic symptoms and prompt activation of public-health protocols. (The possibility of terrorists or other extremists using chemical or biological agents is addressed in Chapter 4, Human-Caused Hazards and Hybrid Hazards.)

## Storm Surges

Storm surges are moving walls of water pushed in front of large weather formations such as tropical storms. While associated with such storms, they are considered a different hazard due to the fact that they are not directly dependent on the winds or rainfall associated with the storm itself, but rather on the body of water over which they travel. Like tsunamis, storm surges do little damage at sea, but that changes drastically when they come in contact with a landmass. Recall that the pumps and levees in New Orleans were managing to handle the intense rainfall of Katrina—until the storm surge emerged from the Gulf of Mexico.

The true threat posed by a storm surge depends on a wide variety of factors. A long, shallow transition from a large body of water onto the surrounding land can allow a surge to build up intensity over an extended front. Storm surges can also be intensified by rocky outcroppings or peninsulas protruding into the sea. Such formations may serve to "steer" a larger volume of water toward and inlet or shallow beach. The pressure, size, and forward speed of a storm will also help determine just how much energy is pushing the wall of water as it approaches the shore.

Finally, the timing of the tide can also be crucial. If a storm surge arrives at the same time as a high tide approaches a shoreline, they can serve as force multipliers for one another, radically increasing the height of the incoming water. By contrast, a low or ebb tide may help weaken an incoming storm surge. Inland lakes and seas are subject to a rare but similar phenomenon known as a seiche, which is essentially a standing wave oscillating in a body of water. Think of it as a bowl of water being rocked back and forth rhythmically, eventually over spilling the lip on one side or the other. Such an effect can be set in motion by specific wind conditions, or even by distant earthquakes.

Densely populated seaboards, such as the Atlantic in America, or the Pacific in much of Asia, lie directly in the path of potential "super surges." Once again, if sea-level rise continues to progress, such areas will be all the more prone to severe storm-surge incidents. Just as with the hurricanes with which storm surges are so often connected, emergency managers must rely on education, early warning, and evacuation to move human populations out of harm's way.

## Tornadoes

Like hurricanes, tornadoes emerge from preexisting weather systems as cyclonic storms of great intensity. The vortex of a funnel cloud is a familiar and terrifying sight to many individuals from the Midwestern United States. "Tornado alley" includes the states of Texas, Oklahoma, Kansas, Arkansas, and Missouri.

Unlike a hurricane, tornadoes are difficult to accurately predict and track, although meteorologists are making progress. They are also much more common than tropical storms, with 1000 or more confirmed sightings annually.

Thunderstorms producing tornadoes occur most commonly over North America, but the phenomena of a funnel cloud have been spotted on multiple continents, resulting in multiple mass-fatality incidents in India, for example. Sustained winds have been measured at more than 300 miles per hour. Most funnel clouds remain aloft, spinning ominously but harmlessly about the surface before getting sucked back up into the storm front from which they emerged. Unfortunately, this is not always the case.

Outbreaks of tornadic activity have killed in excess of 100 Americans on more than a dozen occasions. When a particularly strong line of thunderstorms, often associated with a cold front, marches across the plains states, it is capable of producing clusters or "swarms" of tornadoes. A single cumulonimbus thunderhead could spawn three or more simultaneous funnel clouds, each one a direct threat to lives and property. In the early part of the 21st century, this weather behavior has also become increasingly commonplace in the American South. Current research by Columbia University reveals that the occurrence of multiple-funnel swarm outbreaks doubled between 1965 and 2015 (Thompson, 2016) (Table 3.2).

The intensity of such vortexes can best be measured by comparing wind speeds using the Enhanced Fujita (or Fujita-Pearson) scale. It ranges from 0 (minimal rotation) to 5 (catastrophic):

- EF-0, 65–85 miles per hour
- EF-1, 86–110 miles per hour
- EF-2, 111–135 miles per hour

**Table 3.2** List of the Deadliest Tornadoes in US History

| Date | Location | Fatalities |
|---|---|---|
| **1.** March 18, 1925 | MO, IL, IN | 689 |
| **2.** May 6, 1840 | Natchez, MS | 317 |
| **3.** May 27, 1896 | St. Louis, MO | 255 |
| **4.** April 5, 1936 | Tupelo, MS | 216 |
| **5.** April 6, 1936 | Gainesville, GA | 203 |
| **6.** April 9, 1947 | Woodward, OK | 181 |
| **7.** May 22, 2011 | Joplin, MO | 162 |
| **8.** April 24, 1980 | LA, MS | 143 |
| **9.** June 12, 1899 | New Richmond, WI | 117 |
| **10.** June 8, 1953 | Flint, MI | 115 |

*Source: NOAA.*

- EF-3, 136–165 miles per hour
- EF-4, 166–200 miles per hour
- EF-5, 200 or more miles per hour

Damage from an EF-4 or EF-5 tornado can be extreme, with entire towns or neighborhoods literally leveled by the wind and debris cloud. The majority of fatalities caused by tornadoes are due to structural collapses or by flying debris. Search-and-rescue operations following such storms can last for days, with the devastation delaying access to the impacted area and making the search process itself far more challenging. The recovery phase is no easier.

Advanced warning times and accuracy are increasing, thanks to satellite imagery and improvements such as Doppler radar technology that allows meteorologists to detect rotational motion even within rain-obscured squall lines. Once again, mitigation via stronger structures and better mass-notification procedures is saving lives. Responders should also be training for operating within the hostile and debris-laden environment that will face them following an extreme wind event.

## Tsunamis

As indicated by the disaster cited in the introduction to this chapter, tsunamis are indeed one of the most powerful and least understood threats facing emergency managers.

Any large-scale disturbance below the surface or on the edges of a large body of water carries the risk of producing a tsunami. Tsunamis are by far most commonly caused by earthquakes, but volcanic activity or landslides (above or below the surface) may also be a cause. A tsunami is essentially a wave form emanating outward through deep water from its point(s) of origin. Tsunamis are thought to be capable of traveling across great distances at speeds up to 600 miles per hour, or roughly the top speed of a modern jet airliner. As these waves approach land, they essentially are compressed in frequency and intensity.

This rolling wave motion is responsible for the telltale withdrawal or retreat of water away from the shore just before a tsunami strikes a shallow section of coastline. When the tsunami does arrive, it will appear as a series of massive waves, alternating with troughs as it emerges from deeper water onto the shore. Much like a storm surge, the height, speed, and intensity of a tsunami will depend greatly on the characteristics of where it makes contact with land—particularly the tidal status. Tsunami waves exceeding 100 feet in height have been recorded.

Clearly, such force and magnitude will cause massive damage to anything on or near the shoreline. Most fatalities are due to drowning or debris impact. The recovery and cleanup process for a tsunami is considered extreme, even by disaster standards. A tsunami can be expected to bury formally dry land for hundreds of feet inland in sand, mud, and debris. Such a debris field can stretch inland for miles, depending on conditions, and may permanently alter the shape and geography of a given shoreline.

Tsunami warning systems have improved significantly since the Indonesian disaster of 2004. However, they still leave much to be desired as there are gaps in the capabilities of scattered sensor buoys and the satellite data currently monitoring for the threat. Seismic sensors continue to provide an additional layer of warning for predicting when conditions are favorable for tsunami formation. Ongoing US efforts are led by the National Oceanic and Atmospheric Administration (NOAA).

Much as with hurricanes and storm surges, emergency managers in coastal areas must place efforts into education, early warning, and evacuation to move human populations away from the coastline. However, tsunamis will offer must less advanced notice of their approach than will a hurricane, so public-safety officials must be absolutely prepared to act without delay.

## Wildfires

Wildland fires are considered to be those that are vegetation-based—forests, grasslands, etc.—and are generally geographically removed from significant human development. However, as man encroaches on former wilderness, such a distinct line is becoming more difficult to draw.

There are three traditional categories of wildland fires, as defined by the National Forestry Service: surface fire, ground fire, and crown fire. A surface fire is generally the most common and is what most people probably envision when they picture a forest fire. It moves relatively slowly, and as the name implies, burns primarily at the surface. A ground fire is similar, but may involve underbrush and burn below the visible surface. A crown fir is more fast-moving, potentially leaping from treetop to treetop. All three can involve dense smoke, pockets of intense heat, and the potential to grow larger and more intense.

Increasingly, another type of fire is becoming more commonplace. An intermix or interface fire is one involving both a wildland element and a developed element such as structures or vehicles. In July of 2015, a brush fire driven by strong winds in southern California suddenly blew across Interstate 15, where cars and trucks had slowed to a crawl due to smoke, destroying 20 vehicles and sending many startled drivers fleeing on foot.

Federal resources coordinated by the National Interagency Fire Center in Boise, Idaho, oversee the wildland firefighting effort in the United States. Yet intermix/interface fires may also involve local, tribal, or state responders and resources—which also makes the coordination and chain of command more challenging.

The threat of wildfires impacting urban areas is by no means limited to the United States. In May of 2016, the Canadian town of Fort McMurray essentially evacuated en masse before a massive blaze that locals dubbed "the beast." More than 80,000 residents of the Alberta community were forced to flee, and a full 10% of the structures within the jurisdiction were burned to the ground (Ramsay, 2016, p. 1). A wildland fire forcing the emergency evacuation of a sizable settlement in western Canada would have been virtually inconceivable just a few decades ago, and yet this is a hazard that must now be given serious consideration by urban emergency managers.

As noted previously, the emergency-management community, and its incident command systems in particular, owe a debt of gratitude to the wildland fire community. While not a traditional area of expertise for urban emergency managers, it is imperative that they build knowledge regarding this hazard. Persistent drought conditions, in the western United States and Australia, for example, further contribute to the increasing likelihood of such incidents encroaching into nontraditional jurisdictions.

## Other Natural Hazards

### *Drought*

Drought is a sustained lack of precipitation leading to a shortage of ground water. Major and severe droughts have impacted North America and Africa in the beginning of the 21st century. Considered somewhat of a "stealth" hazard due to difficult-to-define start and end points, drought can nonetheless have a severe impact on a given geographical region.

Placing a strain on potable water supplies can stress any society. Activities from farming to firefighting will be affected by a lack of water. Indirect impacts such as subsidence, increased fire risk, and dust storms may impact transportation and other infrastructure.

### *Thunderstorms*

Thunderstorms are very common events across multiple continents, particularly North America. These events are precipitation-producing cells guided by uneven atmospheric pressure. They frequently include lightning, high winds, and hail.

Thunderstorms are also frequently precursors to related natural hazards such as flooding and tornadoes. Meteorologists have made great strides in predicting the course and severity of such weather events, yet they still remain a threat to people, housing, and infrastructure.

### *Volcanic Activity*

A volcano is a structural vent leading to a reservoir of molten rock or magma below the surface. Taking the form of a conical-shaped hill or mountain over time, they can spew lava, ash, or superheated gas onto the surrounding terrain. Volcanoes are most frequently found near faults or separations in the earth's tectonic plates.

Volcanoes need not be located immediately adjacent to an urban area to cause harm. The eruption of the Eyjafjallajökull volcano on Iceland in 2010 caused a major disruption to air travel in urban centers across much of Europe.

### *Winter Storms*

Many urban areas in North America and Europe also face the prospect of large, seasonal winter storms. Heavy concentrations of snow or ice can snarl transportation and damage crucial infrastructure including power grids.

The primary factors associated with extreme winter weather are severe cold temperatures, high winds, and freezing precipitation. Blizzard conditions are considered to exist when high winds and driving precipitation limit visibility. Travel or emergency response in such conditions can be incredibly treacherous.

## THE URBAN PERSPECTIVE

There are indeed a wide and intimidating range of hazards provided by the natural environment. Each city or urban area will be subject to some, and largely immune from others. It is the responsibility of an emergency manager to assemble as much information as possible about the hazards and threats facing their organization or jurisdiction.

This process is best accomplished via a thorough and honest risk assessment, including natural, human-caused, and hybrid hazards. Additional socioeconomic risk factors will also have a significant impact to a specific urban area's overall hazard status. (The process of risk assessment is addressed in Chapter 5, Assessing Risk in Urban Areas.)

As was noted in the discussion regarding several hazards, preparedness and mitigation measures can be instrumental to an emergency manager hoping to minimize the impact of a specific danger to his or her community. The response phase to a natural disaster in an urban environment is often chaotic and extremely challenging in terms of coordination, timing, and logistics. The recovery phase can be significantly shortened and minimized by effective actions within the other phases of emergency management. The time to assemble an effective emergency operations plan is clearly prior to such an incident occurring.

The back-to-back devastation wrought by Hurricane Harvey across Texas and Hurricane Irma across Florida in August and September of 2017 serves as a reminder of the continuing vulnerability of major population centers to natural disasters. The intense, slow-moving Harvey dumped 50 in. of rain in some areas, leaving nearly 2 million civilians under mandatory or voluntary evacuation orders. Low-lying Houston—a sprawling city with the outdated water-management system common in many American municipalities—and its surrounding suburbs bore the brunt of the resulting and inevitable flooding. In addition to the hundreds of thousands of displaced residents, the storm took multiple large petroleum refineries offline, caused explosions at an organic peroxide plant, and released a range of toxic chemicals into the floodwaters, and into the mud and muck left behind as they finally receded. By contrast, damage from the fast-moving Irma was more related to storm surge and wind. Yet Miami suffered from many of the same urban design shortcomings seen in Houston. In both cases the long-term impact on housing, transportation, and other infrastructure is just beginning to be fully realized. The recovery process for both storms will stretch across several years, with a combined pricetag certain to exceed that of Hurricane Katrina 12 years prior.

No two cities are the same, nor will their hazard profiles remain totally static. A 2015 United Nations report based on historical data indicated that there was nearly one weather-related natural disaster taking place every day in the world, a figure that has nearly doubled over the past two decades (Miles, 2015, p. 2). In a world with an ever-changing climate and hazard profile, the emergency manager will become an increasingly important piece of a truly effective public-safety community.

## SUMMARY

As was proven by the Indian Ocean Tsunami of 2004, one of the greatest disasters in modern history, mankind has in no way fully tamed nature. A wide range of hazards continue to threaten, directly or indirectly, human settlement across the globe. These include, but are not limited to, earthquakes, floods, hurricanes, pandemics, storm surges, tornadoes, tsunamis, and wildfires.

Effective management of such incidents can occur only after detailed and forthright assessments of a given area's natural hazards. Understanding the hazard, and the associated risk factors, is crucial to developing an effective emergency-management plan across all phases of operation.

## References

Haddow, G., Bullock, J., & Coppola, D. (2008). *Introduction to emergency management* (Third edition). Burlington, MA: Elsevier.

Miles, T. (2015). *UN: Weather-related disasters occur almost daily. Reuters.* Available from: <http://www.reuters.com/article/us-climatechange-disasters-idUSKBN0TC1EG20151123>

Ramsay, C. (2016). Fort McMurray wildfire: Federal government commits $300M in disaster recovery funds to Alberta. *Global News*, June 17, 2016.

Thompson, A. (2016). Extreme tornado outbreaks are becoming more extreme. *Climate Central*, reprinted in *Scientific American*. December 1, 2016.

United Nations. (2016). *How do people die in disasters and what can be done?* Available from: <http://disasterdoc.org>

## Further Reading

Birkland, T. A. (2006). *Lessons of disaster: Policy change after catastrophic events*. Washington, D.C.: Georgetown University Press.

Sylves, R. (2008). *Disaster policy & politics: Emergency management and homeland security*. Washington, D.C.: CQ Press.

CHAPTER 4

# Human-Caused Hazards and Hybrid Hazards

## INTRODUCTION

On December 2, 1984, a large Union Carbide chemical facility in Bhopal, India, developed a significant leak. The plant, a joint venture between the US corporation and Indian government interests, used various chemicals to produce the pesticide carbaryl. On the evening in question, water inadvertently entered a large holding tank, releasing a massive cloud of methyl isocyanate (MIC) onto the surrounding, highly populated terrain during nighttime and early-morning hours.

The results are well known to emergency-management professionals, the specific causations less so. Crucially, the chemical reaction in the holding tank was magnified by a number of malfunctioning or missing safety valves and by the relatively high temperatures in the region. Over the next several hours, the surrounding shanty towns—many of which were occupied by plant employees and their families—were slowly and silently overrun by a cloud of pesticide. Many perished in their sleep, while others fell dead as they attempted to escape on foot in the darkness.

Both the local medical system and the state government of Madhya Pradesh were overwhelmed by the number of casualties and the scope of the disaster. Information released to the public was scant, poorly coordinated, and delayed. The official death toll was soon placed at 2259 but more reliable estimates range from 5000 to 8000 fatalities, with up to 500,000 additional inhalation injuries. Tens of thousands of livestock were also thought to have been killed by the toxins, and agricultural fields in the area lay fallow for years for fear of residual contaminants in the water and soil.

The events in Bhopal represent a particularly stark and frightening technological example of a human-caused disaster. Numerous opportunities to prevent or mitigate the disaster had been squandered, and the results were catastrophic.



Urban Emergency Management. DOI: http://dx.doi.org/10.1016/B978-0-12-805307-2.00004-4

## ANALYSIS

No other species has altered the planet to meet its needs quite like Homo Sapiens. And the vast majority of the time, things go our way. The engineered infrastructure we design is successful. The large-scale agricultural project increases crop yields. The reallocation of natural resources goes reasonably well. The individual planning to cause harm to others has second thoughts or is interdicted by law enforcement.

But sometimes, rarely, things do not go our way. Infrastructure fails. The mismanagement of resources leads to civil unrest. Terrorists slip through. And the results are human-caused disasters; manmade catastrophes that we essentially inflict on ourselves.

### Risk Factors

As noted in the discussion of natural hazards, a number of recognized risk factors are at play in the severity of any human-caused hazard. Again, these may be seen as magnifiers or multipliers, exaggerating positive or negative impacts of a given hazard for certain groups or subsections of the larger population. Such social, cultural, and economic traits are referred to as risk factors.

In reference to our earlier example of the chemical release in Bhopal, the number of fatalities was greatly increased by the magnified risk factors of a largely poor and illiterate populace in close proximity to the site of the incident. The population's lack of warning systems and vehicular mobility were also contributing risk factors to the final outcome. As was noted for natural disasters, in dense urban environments, these advantages or disadvantages also tend to be quite geographic in nature, often dividing neighborhoods along economic lines.

Recognized risk factors include:

- Age
- Gender
- Race
- Religion
- Health
- Security
- Preparedness
- Literacy
- Wealth
- Credit
- Savings
- Insurance

## TYPES OF HUMAN-CAUSED HAZARDS

A human-caused hazard may be defined as a source or cause of potential loss with mankind as its primary driver. The engineered world is increasingly complex, and the types of hazards it is capable of generating varies widely by geography, type, and severity. The significant or more frequently occurring intentional or unintentional instances are worthy of further review.

### Active Threat

It is an unfortunate fact that emergency managers may be called upon to assist with active-threat incidents, particularly in the United States. An active threat is a situation in which an armed individual is uncontained and a direct threat to lives and property. While ultimately a security and law-enforcement issue, such situations have the potential to benefit from emergency-management involvement across all five phases.

Unlike a hostage-barricade situation, which is static, an active threat involves a perpetrator or perpetrators actively attempting to injure or kill victims. Firearms are generally the weapons of choice, although blades and blunt objects have also been used in previous incidents. And while there could be a nexus to terrorism, active threats can also be the result of mental imbalance, workplace disagreements, or domestic violence.

In terms of human-caused hazards, active threat or active shooter situations receive a very high level of media coverage, and thus extreme public awareness and sensitivity. Based on studies of recent incidents, most active shooter scenarios last between 5 and 12 minutes. The arrival of trained security or law-enforcement personnel frequently marks the end of the threat, and roughly half of all perpetrators take their own lives when confronted.

Emergency managers can be especially effective when combining their expertise with law-enforcement and security professionals, particularly in the preparedness and mitigation phases. Training civilians within their jurisdiction, or employees within their organization, emergency managers can make these groups far more prepared for any possible active threat scenarios. It is a telling commentary on our times that the Department of Homeland Security's (DHS's) "Run, Hide, Fight" maxim is becoming as commonplace as the "Stop, Drop, Roll" fire guidance made famous decades before. However, the importance of preparation and planning cannot be overstated.

If the prevention phase fails and an attack does take place, anticipate the incident scene to be tense and emotional. In the response and recovery phases of an active threat, emergency managers may be called upon to perform a wide range of duties, ranging from liaison to mortuary services to coordinating reentry credentials.

## Civil Unrest

Alongside the concentration of residents in urban centers comes the possibility of civil unrest. Civil unrest may be viewed as a disruption or disturbance evolving from political, social, or economic causes. Symptoms may include traffic disruption, melees with security or law-enforcement personnel, vandalism, and looting.

This may, in its most mild form, take the form of a small and peaceful group of marchers, obtaining permits and permission in advance, and moving in coordination with local law enforcement. At the other end of the spectrum, it may involve spontaneous, violent, and enormously large-scale demonstrations.

Note that not all civil unrest begins as a form of political protest. Some of the most damaging unrest begins when sports fans spill into the streets to celebrate a team's major victory, or to mourn its loss. Group psychological dynamics, and often alcohol, plays a huge role in such crowds remaining peaceful or "breaking bad" and causing injuries and property damage. Veteran law-enforcement professionals will swear that they can sense which sets of behavior a crowd is going to display well in advance.

It is absolutely crucial that emergency managers in every sector be involved and aware in planning for such events. A truly spontaneous round of civil unrest is extremely rare. Much more often, such instances evolve out of a planned march or at the conclusion of a league championship match. Having an awareness of planned events within one's jurisdiction or area of responsibility is therefore vital. Situational awareness and coordination with fellow public-safety professionals is the best course of action for a well-prepared urban emergency manager.

## Cybersecurity

In an increasingly interconnected world, cybersecurity is becoming a critical factor to the success (or failure) of any jurisdiction or organization. Badly written or obsolete computer code can cause widespread problems, but the hazard analysis here will focus on intentional theft, harm, or disruption via computer-operated systems.

Theft is indeed the most common goal of so-called computer hackers with malicious intent, sometimes referred to as cybercrime. Very frequently, this crime targets an established government or corporation and is perpetrated by a developing nation or financial competitor. At a security conference in 2012, Michael Hayden, the former director of both the American Central Intelligence Agency (CIA) and the National Security Agency (NSA), called such electronic espionage "the greatest theft of intellectual property in human history."

But exfiltration of data is not always the intent. A December, 2015, power outage in Ukraine was the first widely acknowledged disruption of civilian infrastructure in the physical world by an attack originating in the cyber world. (A number of previous incidents were disproven or traced back to unintentional failures.) In the 2015 operation, malicious code originating in Russia was intentionally fed into the supervisory control and data acquisition (SCADA) system operating a portion of the electrical grid. The resulting outage resisted traditional "reset" mechanisms due to the nature of the malware. (See the brief case study in Chapter 9, Prevention Phase.)

This latter instance—a direct assault on critical infrastructure—represents a smaller but more immediate threat to emergency managers. Whether originated by state actors, malicious hackers, or cyber terrorists, such attacks can have a real impact on the command and control systems key to emergency managers.

The electrical power grid is an excellent example. According to an assessment by the *Washington Post*, in the United States alone there are 7700 power plants, 55,800 substations, and 700,000 miles of primary, high-voltage transmission lines (Mooney, 2016). The federal government has acknowledged that the rapidly evolving system is becoming more susceptible to external cyber attacks, not less. And Purpura notes that not just power grids are vulnerable but also financial institutions, air-traffic control systems, water-management systems, and other vital technologies (2007, p. 61).

Further challenges are posed by the prevalence of outsourcing so-called cloud storage—usually to the lowest bidding provider of server storage space—and by the expansion of complex encryption applications for communications and handheld devices. Such rapidly evolving technologies can ultimately be double-edged swords, proving extremely helpful and useful when in the right hands, and extremely destabilizing and threatening in the wrong hands.

Emergency managers should always follow established electronic-security protocols, and must work closely with their technical services teams to be certain that their critical devices and applications are up to the latest standards. The most common exploit utilized by malicious actors is outdated or "unpatched" systems or applications, as evidenced by several widespread and high profile "ransomware" attacks in 2017.

Redundancies and back-up systems should be the order of the day, as the ability to operate in austere environments under technologically challenging conditions—such as a temporary inability to access cellular service or the Internet—should always be considered a likelihood. At the same time, managers cannot be afraid of taking a chance on a new technology that may make their lives easier, particularly during disasters. Just consider the revolutionary leaps forward provided by technologies such as reliable cellular service, satellite communications/positioning, and Doppler radar systems.

The response and recovery phases of a cybersecurity incident will likely require effective emergency managers to display skills in information technology, infrastructure awareness, and communications.

## Hazardous Materials Releases

Hazardous materials are substances that, if released or mixed with other elements, can pose risks to health, infrastructure, and the environment. When contained with proper precautions, such chemical or biological substances are perfectly safe. Yet when containment is lost and the material comes in contact with flame, air, electricity, air, water, or other substances, the reaction can be immediate and sometimes violent. Emergency managers often refer to these substances as hazmat. In modern urban environments, the most common applications for such materials include refrigeration, transportation, medicine, research, cleaning and sterilization, and industrial purposes (Table 4.1).

**Table 4.1** Indicators of Possible Chemical Incident

| Indicator | Description |
| --- | --- |
| Dead animals/birds/fish | Not merely occasional road kill, but numerous animals (wild and/or domestic), birds, and fish in the same area |
| Lack of insect life | If activity lacking, examine ground/water surface/shoreline for dead insects. If near water, check for dead fish or aquatic birds |
| Unexplained odors | Smells may range from fruity/flowery to sharp/pungent to fresh hay to garlic or almond-like. Important to note if odor is out of character with its surroundings |
| Large number of victims/ mass casualties | Health problems may include nausea, disorientation, breathing difficulty, convulsions, excessive sweating, reddening of skin/eyes, or death |
| Pattern of casualties | Casualties will likely be grouped together, downwind, or near ventilation outlets |
| Blisters/rashes | Widespread symptoms of visible blisters, weals, or rashes |
| Illness in confined area | Differing casualty rates and severities for indoor versus outdoor victims |
| Unusual liquid droplets | Numerous surfaces, potentially including standing water, exhibiting oily droplets, and/or film |
| Differing vegetation | Not just a patch of dead weeds, rather trees, shrubs, bushes, crops, and/or lawns that are dead, discolored, or withered |
| Low-lying clouds/haze | Low-lying cloud/fog-like conditions that are not consistent with surrounding weather |
| Unusual metal debris | Unexplained bomb or munitions-like material, especially if liquids are present |

*Source: Emergency Response Guidebook.*

The carbaryl pesticide formerly manufactured in Bhopal serves as a prime example. When contained in a properly pressurized and sealed tank, the substance was comparatively harmless. But when exposed to water and allowed to vent, the gaseous byproduct was uncontained and highly toxic.

Industrial quantities of hazardous materials are stored or utilized at more than 4.5 million facilities in the United States alone (Haddow, Bullock, & Coppola, 2008, p. 53). This may range from a few bags of nitrogen-based fertilizer at the local gardening store to thousands of barrels of highly toxic waste produced by a nearby metal-finishing plant. Or, as Erickson notes, each of the roughly 60,000 chemicals in daily commercial use in developed nations could be considered hazardous to humans under the wrong circumstances (2006, p. 105). As a result, most nations and jurisdictions have strict guidelines requiring signage or placards to be visible on buildings or transportation vehicles where hazmat is present. Many urban fire departments have specially trained and equipment hazardous materials units to respond to incidents at such facilities, or incidents involving unknown substances (Fig. 4.1).

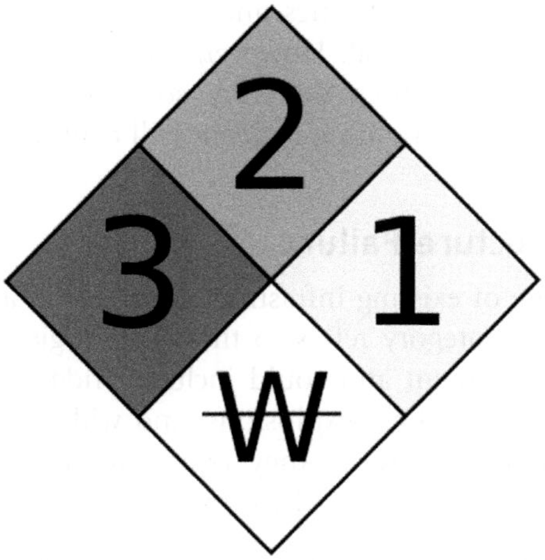


**FIGURE 4.1**
The universal hazardous materials diamond marking designates the health hazard, flammability, instability, and special characteristics of a specific location.

A few key terms and measures explain just how dangerous a given material may be. Relative toxicity refers to the differences between substances in terms of the quantity required to cause death in the majority of those exposed. Some chemicals require exceedingly small amounts—only a few milligrams—to make fatality highly likely. Such substances are generally referred to as extremely toxic or poisonous.

The route of entry or means of contact will also have a significant influence on the victim's reaction to a hazardous material. The primary methods are:

- Inhalation, breathing in via the nose or mouth
- Ingestion, eating, or drinking resulting in exposure via the digestive tract
- Absorption, entry via the skin, eyes, or other membranes
- Injection, exposure via penetration injury or hypodermic

Because of their ubiquitous nature, hazardous materials represent a unique risk to urban emergency managers. Specialized fire-services or public-health personnel will likely be the primary responders to hazmat events, utilizing sensitive detection devices and personal protective equipment (PPE). However, emergency managers must place efforts into planning and education, particularly for employees in industrial or transportation settings. Educating the general public about shelter-in-place and evacuation procedures is also vital to minimizing injuries and fatalities during hazmat releases. Logistical abilities and knowledge of hazardous materials or basic chemistry will be vital. However, the symptoms and severity of exposure to dangerous substances can vary so widely that effective recognition and response to an unfolding incident will continue to be critical to the outcome.

## Infrastructure Failure

The failure of existing infrastructure is always a concern in modern societies. This broad category refers to the technological hazards that result from the built environment and could include bridge or dam collapses, construction accidents, natural-gas explosions, and widespread power outages, to name a few. Taken as a group, they represent the greatest likelihood for a mass-casualty incident in virtually every urban jurisdiction.

Critical infrastructure is a system of interdependent nodes and networks that support the functionality of multiple sectors of the economy and society as a whole. Commonly cited elements may include the electrical grid, natural-gas lines, and telephone networks. Less recognized elements may include cellular networks, transportation systems, and financial institutions. In the modern and interconnected world, a failure or degradation of one subsystem may ripple across the larger network to impact others.

The failure of the Interstate 35 bridge in Minnesota in August of 2007 provides a textbook example of an infrastructure failure. The span over the Mississippi, built in the mid-1960s, was undergoing extensive reconstruction when it suddenly collapsed, killing 13 and injuring scores more. The bridge was being repaved and retrofitted, and more than 500,000 pounds of sand and construction equipment were present at the time of collapse. Engineers had previously cited concerns about the bridge's structural integrity, some of which were to have been addressed in the ongoing renovation project. Undersized gusset plates—key components in the structure's ability to absorb weight and lateral force—were cited as the primary point of failure.

The cascading impacts of such a catastrophic engineering failure are not always immediately evident. Communications and fiber-optic cables are often strung beneath bridges, as are natural gas and petroleum pipelines. The I-35 collapse also completely blocked the Upper Mississippi River for several days, one of the busiest bulk-shipping waterways in North America. It is this interconnectedness of critical infrastructure that magnifies the impact of any such failure in an urban setting. An emergency manager's skills in infrastructure, public works, engineering, and search and rescue may come into play following a major failure.

From an emergency-management perspective, knowledge is truly power. Managers in the public sector must know where the critical infrastructure is located in their jurisdiction, and how it interconnects with other nodes and subsystems. Similarly, managers in the private and nonprofit sectors must know what external systems are critical to their organization's operation, and which internal subsystems are equally vital. This unquestionably requires forethought and communications with other vital stakeholders, including your organization or agency's internal experts, utilities, key vendors, and—in the United States—intelligence fusion centers.

Such information gathering will require substantial effort and coordination. However, recognizing a key infrastructure component only after it has been put at risk is a fully avoidable mistake. With knowledge comes the ability to develop redundancies, consult subject-matter experts, and identify effective prevention and mitigation strategies.

## Nuclear Incidents

While such incidents could be considered a subset of the larger infrastructure or technological failure category, the challenges presented by a radioactive release are so unique as to demand separate analysis.

Perhaps nothing strikes fear into the general public quite like a radiological event. It must be stated that nuclear fission has, overall, been an incredibly safe and highly regulated field of the larger utilities industry. Yet from Three Mile Island in the United States, to Chernobyl in the former Soviet Union, to Fukushima in Japan, nuclear energy does have the ability to embody mankind's worst fears.

The Chernobyl disaster in April of 1986 provides a well-documented example of this subtype of human-caused hazard. A power surge led to steam explosions and a serious fire within one of the primary reactors at the facility, located near the town of Pripyat, in modern-day Ukraine. A portion of the nuclear core became exposed and an uncontrolled nuclear reaction began—commonly referred to as a "meltdown." Thirty-one immediate deaths resulted, including plant workers and emergency responders. Thousands of cases of cancer, many fatal, have been subsequently tied to the mass radiation release that followed. A cloud of highly radioactive particles drifted across the Soviet Union as well as portions of eastern and northern Europe. Soviet authorities were typically tight-lipped about the extent of the disaster, and much of the actual radiation measurement was instituted and publicized by Western nations.

Herein lies the danger of nuclear power. The radiation itself is essentially a colorless, odorless series of waves and particles traveling through the air. Without adequate containment and warning systems, civilians thousands of miles away from a given incident could be unknowingly exposed to dangerous levels.

Again, emergency managers will be relying on subject-matter experts to inform their planning and actions. Foreknowledge is especially critical in jurisdictions or facilities with reactors present, whether large energy-producing models or the smaller research and medical units found in many Western academic institutions. Radiation protocols again must focus on evacuation and shelter-in-place planning, allowing specialized responders the time and space needed to regain containment and mitigate the threat. Key guidance is available via the Nuclear Regulatory Commission (NRC) in the United States, and by the Western European Nuclear Regulators Association across much of Europe.

## Structure Fires

According to the United States Fire Administration (USFA), there are over 1.5 million fires reported to responding agencies in the United States annually. This number is divided almost evenly between outdoor or wildland fires and fires occurring in structures or enclosed spaces. The former are generally considered to be natural, and the latter human-caused.

This is not to say that lightning does not sometimes cause house fires, or that forest fires are never caused by human error. Rather, it acknowledges the systemic differences in fighting a wildland fire and a structure fire. This differentiation is thus based as much upon multiphase emergency-management tactics as upon causation.

Structure fires are likely to be the single most common incident type faced by an urban emergency manager. Cities provide a high density of buildings and human population, and a wide variety of ignition sources. Note that fires can also be a cascading or secondary disaster type, resulting from primary causes ranging from earthquakes to severe weather to terrorism.

All cities have professional firefighting departments, yet the training and quality of these first responders varies widely. Many urban facilities, particularly high rises, have residents or tenants assigned to fire warden or safety warden roles. This is particularly true within corporations or other centralized organizations that have safety programs in place for concentrated numbers of employees or associates. Working closely with those who will be executing prevention and response efforts on your behalf is the responsibility of any effective crisis manager.

During the response phase, an emergency manager will not be donning bunker gear and an air tank to rush into a burning building. However, he or she will have to deal with key questions regarding the consequences of the fire, and all tangential impacts. When are search and rescue efforts complete? Were residents or businesses displaced? Are mortuary services required? Has adjacent infrastructure been damaged? How will access again be granted once the area is deemed safe and secure by the incident commander?

As a professional emergency manager in an urban environment, you will encounter large and complex structure fires. The outcome of these incidents will largely depend on the quality of the training provided to both civilian and professional responders. A high-performing emergency manager can have a proactive and meaningful impact by collaborating with multiple parties across varied sectors to inform and guide prevention and mitigation efforts before a fire ever occurs, and by demonstrating keen logistics, infrastructure, and communications skills during the response and recovery phases.

## Terrorism

No other threat or hazard has captured the public's attention in recent years as has that of terrorism. Large-scale attacks have spanned multiple continents in the new millennium, as have multiple smaller-scale incidents, particularly in cities across Europe and the Middle East.

Terrorism has a range of definitions, but all include violence or the threat of violence by one group upon another based on political or ideological goals.

It is not a new phenomenon, nor will it be one that disappears in the foreseeable future. And emergency managers are a part of the front line of defense in urban areas.

Modern terrorist organizations defy recognized national boundaries and are constantly splintering or forming new alliances. The traditional categories of terrorism are domestic and international (reflecting the scope or scale of the objectives of a given terror group), or state and nonstate (reflecting the primary sponsor or funder of a given terror group). Modern extremist organizations may not easily conform to such well-defined "boxes." The tactics or methodologies utilized by such groups also continue to evolve, but include:

- Assassinations or targeted killings
- Bombs, incendiary, or explosive devices
- Chemical, biological, or radiological releases
- Cyber attack
- Hijacking
- Kidnapping or hostage-taking
- Sabotage
- Small-unit or group assault
- Vehicle as a weapon

Each type or tactic represents a unique set of challenges to emergency-management and homeland-security professionals. (These roles are further discussed in Chapter 8, Emergency Management and Homeland Security in Urban Areas.) In essence, first responders should presume an incident is terrorist-related until proven otherwise.

### *US Federal Guidance*

Managing terrorism involves a wide range of activities involving planning, training, and preparedness. In the United States, since the September 11 attacks, one of the primary documents guiding these "big picture" responsibilities is the National Response Framework (NRF). Created in 2008, the NRF is itself an evolution of several previous documents, most recently the National Response Plan (NRP) of 2004. The National Preparedness System (NPS), introduced in late 2011 and revised several times since, builds on the overall planning and preparedness model with similar language.

Each of these guides represents a refining of the core principles of emergency response on a national scale and sets forth expectations for interactions between local, state, and federal responders. All address a cycle of high-level goals—prevent, prepare, respond, recover—which represent recurring themes throughout the federal guidelines regarding emergency services. They are referenced in planning cycles, in objectives, in strategies, and in tactics.

The NRF also lays out a National Strategy for Homeland Security. Four key elements comprise the role of the DHS in this function:

1. Prevent and disrupt terrorist attacks
2. Protect the American people and our critical infrastructure and key resources (CIKRs)
3. Respond to and recover from incidents that do occur
4. Continue to strengthen the foundation to ensure our long-term success

As defined in the NRF, DHS must keep these four strategies at the core of its overall mission (2008, p. 12). The first goal or strategy falls primarily to the US military, and to law-enforcement and intelligence services within the federal government, including the CIA, National Security Administration (NSA), and Federal Bureau of Investigation (FBI).

The second goal is closely tied to the first, in that it focuses on preventing and mitigating acts of terrorism against American citizens or interests. A key portion of this strategy will fall to state and local governments, in that they will need to define and identify CIKRs within their jurisdictions. Emergency managers will thus be a key component prior to any terrorist incident and can also help drive resiliency efforts to prompt an efficient recovery.

As noted in its text, the NRF "primarily focuses on the third goal: respond to and recover from incidents that do occur … it is our collective duty to provide the best response possible" (2008, p. 12). This is the portion of the strategy that acknowledges that incidents will occur that require local and state responders—and their partners within the private sector—to present an organized and effective contribution in the response and recovery phases. Again, emergency managers will be an inherent part of such response and recovery efforts.

The fourth and final goal is perhaps more esoteric and cerebral than the proceeding three. It requires a constantly improving system of principles and systems to deal with potential threats to homeland security. These include, but are not limited to, changes in culture, technology, and management.

Also worthy of mention is the Urban Area Security Initiative (UASI), a federally funded program that pushes state and local entities to take a greater role in all phases of emergency management related to homeland security (Sylves, 2008, p. 183). As the acronym implies, UASI is active across the United States in all major urban centers, although there is ongoing funding pressure to reduce that breadth and scope. The program provides grants for equipment and services ranging from security bollards and barriers around sensitive sites, to training sessions and equipment for specialized response units.

### *Emergency-Management Best Practices*

The line between emergency management and homeland security can be blurry or ill-defined. (See Chapter 8, Emergency Management and Homeland Security in Urban Areas, for a more extensive analysis.) When responding to an incident in the field, all public-safety personnel should always presume that the incident is related to terrorism until proven conclusively otherwise.

If a terrorism nexus is suspected, emergency managers and other first responders should initiate a number of key tasks or actions:

- Assess the situation, determine status and types/numbers of casualties
- Be alert for secondary devices or additional suspects
- Identify a "hot zone" or area still under threat
- Establish a perimeter(s)
- Establish a forward command post (FCP) and/or emergency operations center (EOC)
- Identify areas for triage and staging of resources
- Request additional personnel and resources

A veteran responder once told the author that 50% of the information received in the immediate aftermath of an incident will be completely or largely untrue. That may be a generous percentage. Given the "fog of war" that is often present in the minutes following a major incident such as terrorism, rumors and false leads abound. During active shooter situations, witnesses commonly overestimate the number of shooting suspects involved. On the morning of September 11, 2001, multiple agencies and media outlets reported various attack locations—such as US State Department headquarters—that proved to be categorically untrue.

A skilled emergency manager's role, whether in the field or in the EOC, is to filter and decipher a vast number or reports in search of the true status on the ground. In the present environment, emergency managers will be key players across all phases of antiterrorism planning and response.

## Transportation Accidents

Finally, we must consider the likelihood of our interconnected world experiencing a transit interruption. In a complex, fast-moving world, there are many potential transportation issues that we must confront on a daily basis.

Overall, modern transportation systems are amazingly efficient and safe. Incidents rising to the level of the emergency manager are exceedingly rare. And yet the fact remains, trains sometimes derail, buses and planes sometimes crash with dozens or even hundreds of passengers inside. And as

Birkland notes, this rarity and shock can cause them to become defining historical events for an organization or jurisdiction (2006, p. 62).

When such incidents do occur, it will be with little or no warning, under immense media scrutiny, and with the full attention of your top elected officials or senior corporate executives. (To this day, the largest-scale incident to which the author has responded was an Amtrak train derailment with approximately 180 casualties—all of whom survived thanks to the Chicago Fire Department and excellent local hospitals.)

It is also worth noting that transportation disasters often spill over into other types as well, impacting other infrastructure (a bus striking an electrical tower), or creating cascading threats (spilled fuel becoming a hazardous materials release). Such incidents are often complex and uncontained when they first occur.

Preplanning is again a key factor. An effective emergency manager will know the major transportation modes and routes in his or her area. A basic understanding of transportation systems, such as how many flights per day the local airport hosts or the seating capacity of the city's standard passenger bus, is extremely useful knowledge to have.

Corporate or nonprofit crisis managers must know their counterparts in the public sector, and vice versa. A fast-moving emergency such as a transportation incident is not the time an effective manager should be scrolling through smartphone contacts or digging through a Rolodex on the desk; knowing your counterparts is invaluable in such circumstances. (This point will be further addressed in Chapter 7, Collaboration in Urban Environments.)

Finally, the emergency manager may serve as a key liaison between the transportation company/provider and other stakeholders such as local and federal investigators, family members, and the media. Following a large-scale transportation incident, it is highly recommended that a joint information center (JIC), private family gathering point, and EOC be immediately established.

## THE HYBRID DISASTER

It is imperative that emergency managers not be locked into a rigid way of thinking about the threats and hazards they face. One emerging trend worthy of further study is the concept of a hybrid disaster—one that requires a combination of both natural and manmade causes to occur.

As the world—and our understanding of it—becomes a more complex and interconnected place, the hybrid disaster concept is likely to gain further

**FIGURE 4.2**
Complex incidents may require extensive or specialized personal protective equipment such as self-contained hazardous-materials response suits. *Credit: Shutterstock.*

acceptance. Over many millennia of evolution, the planet changed the human species. We became adapted to a wide variety of weather conditions, from the broiling topics to the frigid artic. We adapted our diets to what was available in a given region. We migrated from continent to continent seeking new and fertile living spaces.

And somewhere along the line, we in turn began to change the planet. We terraformed our surroundings by planting edible or useful crops. We domesticated animals. We built machines and cities and roads. And we started powering those machines with petrochemicals. All of which, in turn, began to change our surroundings (Fig. 4.2).

The natural and engineered worlds have now become so intertwined that it is virtually impossible to put an individual disaster in a specific causation "bucket." Was that flood a natural occurrence, or did it have a larger impact because the river system had so many levees, dikes, and human settlements along its banks? Was the earthquake an especially severe one, or were the resulting fatalities due to poor construction quality and building codes near the faultline? The resulting arguments could result in some unpleasant answers, but such questions must be asked.

## Cascading Crisis in Fukushima

Some disasters are truly and undeniably combinations of natural and man-made phenomena. Perhaps the most notable hybrid disaster in modern times took place in Japan in March of 2011. It is a scenario so complex and nuanced that any novel editor would have rejected the plot as too farfetched and unrealistic to ever actually take place.

Just before 3:00 p.m. on March 11, 2011, off the eastern coast of Japan, two tectonic plates shifted violently causing a large offshore earthquake. Known as an undersea megathrust quake, it was similar to the disturbance that caused the catastrophic Indian Ocean Tsunami in 2004. In this case the temblor measured 9.0 on the modified Richter Scale, making it the strongest earthquake to strike Japan in modern history.

Fortunately, the quake was centered far enough offshore that it did only light to moderate damage to mainland Japan, in part due to the nation's quite stringent building codes. Unfortunately, the earthquake was only the first in a related series of three interconnected disasters. Similar to the events in the Indian Ocean a few years prior, the undersea shift generated a massive tsunami.

As the series of massive crests rushed outward from the undersea rift, early warning systems in Japan and in satellites far overhead alerted authorities to waves with the potential to reach up to 40 meters (130 feet) in height. Many Japanese civilians heeded the warnings and headed for higher ground, all hoping that their extensive seawalls would hold back the Pacific Ocean. They did not.

The immediate impact ranged from severe to catastrophic, depending on the offshore topography and the height of the nearby seawall. Entire towns and villages on Japan's east coast were simply swallowed by the sea, resulting in nearly 16,000 immediate fatalities and approximately a quarter-million homeless civilians. More than 45,000 structures were completely destroyed and 25 million tons of debris would be left behind.

And the suffering was not yet done, as the third act began to unfold.

The Fukushima Daiichi nuclear power plant was in the direct path of the oncoming waves, at least one of which topped the nearby seawall by more than 3 meters (9.5 feet). This rush of water flooded the lower levels of the nuclear facility, knocking out key systems including the back-up generators that fed the coolant pumps. When emergency batteries began to fail, the coolant stopped circulating in the damaged reactors. The lack of coolant in turn led to the partial meltdowns of reactors 1, 2, and 3 within a matter of hours.

Nuclear material was released into the air and water, resulting in widespread contamination and the additional evacuation of thousands of nearby residents. Initial containment efforts failed, and the only way to cool down the partially exposed nuclear cores was to douse them with water. More than

5 years after the triple disaster, seeping radiation continues to leak into the Pacific Ocean to be carried far and wide by its currents.

### Lessons Learned

No other recent disaster—or, more accurately, series of disasters—better illustrates the concept of hybrid causation. No reasonable planner could have foreseen a large earthquake, followed by a tsunami, which in turn overtopped the seawalls, which then flooded critical equipment, which finally caused a partial meltdown due to (with bitter irony) a lack of liquid coolant. And yet, that chain of events is precisely what took place. A true hybrid disaster had unfolded right in front of the international media.

And it will not be the last. It is precisely this very global and interconnected world that provided the opportunity for such a disaster to occur and for a flood of international relief, aid, and expertise to follow. Emergency managers must look at the big picture and take into account a cascading series of events, which lead to a magnified outcome.

We must be able to envision an earthquake causing a tsunami causing a nuclear disaster. (Just as we must envision terrorism motivating multiple airliner hijackings causing the collapse of two of the largest buildings in the world.) It is vital that emergency-management professionals be able to conceptualize the full scope of hybrid causation. Rather than always focusing on the worst-case scenario, this thinking must encompass a range of potential impacts and outcomes.

Ultimately, the hybrid disaster is here to stay. This intermixed natural and manmade planet is undergoing an unprecedented shift in climate. The oceans will soon be higher, leading to more cascading impacts due to tropical storms and other offshore events. Droughts will occur where they have not taken place in our lifetimes. Our technologies continue to interact with the natural world in new and sometimes unpredictable ways. And emergency managers must be prepared for as many scenarios as we possibly can—even those that seem too farfetched to believe.

## THE URBAN PERSPECTIVE

Human-caused and hybrid disasters are indeed a varied and threatening lot. Each city or urban area will be subject to some, and largely immune from others. As noted previously, it is the responsibility of an emergency manager to assemble as much information as possible about the hazards and threats facing their respective organization or jurisdiction. This process is best accomplished via a thorough and honest risk assessment, including natural, human-caused, and hybrid hazards.

Foreknowledge, preparedness, and mitigation measures can be instrumental to an emergency manager hoping to minimize the impact of a specific danger to his or her community. The time to assemble an effective emergency operations plan is clearly prior to such an incident.

Technological and manmade hazard profiles for a given city will change over time. Particularly in an urban environment, a connected and proactive emergency manager will become an increasingly important piece of a truly effective public-safety community.

## SUMMARY

A number of human-caused and hybrid disasters were examined, including the chemical release in Bhopal and the earthquake/tsunami/meltdown in Fukushima—places that will forever be associated with those incidents. A wide range of hazard profiles and types were reviewed, including: cybersecurity, hazardous-materials releases, infrastructure failure (including nuclear and transportation incidents), structure fires, and terrorism.

Facing such a broad array of hazards makes it clear that emergency managers must perform detailed and forthright hazard and vulnerability assessments prior to any disaster. Understanding complex or multifaceted hazards will be increasingly crucial to developing effective emergency-management plans in an increasingly complex environment.

## References

Birkland, T. (2006). *Lessons of disaster: Policy changes after catastrophic events*. Washington, D.C: Georgetown University Press.

Erickson, P. (2006). *Emergency response planning for corporate and municipal managers*. Burlington, MA: Butterworth-Heinemann.

Haddow, G., Bullock, J., & Coppola, D. (2008). *Introduction to emergency management*. Burlington, MA: Butterworth-Heinemann.

Mooney, C. (2016). New Obama report warns of changing environment for the electricity grid. *Washington Post*. Available from: <www.washingtonpost.com>.

Purpura, P. (2007). *Terrorism and homeland security*. Burlington, MA: Butterworth-Heinemann.

Sylves, R. (2008). *Disaster policy & politics*. Washington, D.C: CQ Press.

## Further Reading

Hayden, M. (2012). *Remarks to Aspen Security Conference*. Aspen, CO.

Department of Homeland Security. (2008). *National response framework*. Washington, D.C.

Department of Transportation. (2016). *2016 Emergency response guidebook*. Washington, DC.

CHAPTER 5

# Assessing Risk in Urban Areas

## INTRODUCTION

The morning of September 11, 2001 was clear and sunny on the eastern coast of the United States. Just after 8:00 a.m., officials at American and United Airlines, and the Federal Aviation Administration (FAA), were confounded by initial reports of airliners being hijacked. The first reports were confusing, including calls coming from individuals onboard the aircraft claiming to be members of the cabin crew. Airline managers were initially incredulous—after all, an actual hijacking of a US domestic airliner had not happened in decades.

Meanwhile, air traffic controllers, unable to reach the aircraft by radio, watched on their radar screens as multiple airliners began to deviate from their designated flightpaths. They too were unsure of what to think. Was it a communications failure? Had the aircraft turned back due to some mechanical emergency?

And then, at 8:46 a.m., American Airlines Flight 11 struck the North Tower of the World Trade Center.

The airlines and the FAA quickly recognized they were looking at a new kind of threat, one never before seen in the United States. Civilian airliners were being used as guided missiles against targets in the American homeland. Although confusion reigned that morning, the two groups began furiously sharing information and data to accomplish the critical task of sorting out which aircraft were involved, and which were not.

Thomas Kean, Chairman of the National Commission on Terrorist Attacks Upon the United States, commonly known as the 9/11 Commission, would later assess the oft-analyzed situation in the group's benchmark report. "We were unprepared. We did not grasp the magnitude of a threat that had been gathering over a considerable period of time. As we detail in our report, this



Urban Emergency Management. DOI: http://dx.doi.org/10.1016/B978-0-12-805307-2.00005-6

was a failure of policy, management, capability, and, above all, a failure of imagination."

What Mr. Kean ultimately meant is that the attacks were the result of a failure to recognize a series of emerging risks; a failure to perceive the operational capabilities of Al Qaeda, to recognize hijacking to be an ongoing threat, to imagine that airliners could be used in such a manner. Assessing risk is a complex and challenging task that requires imagining the unimaginable. It is also the absolute foundation of effective emergency management.

## ANALYSIS

Risk can be a moving target. Very few individuals had envisioned the use of large civilian aircraft as guided weapons capable of causing enormous damage and loss of life. And yet, in hindsight, the risk of such an occurrence obviously existed.

Here, it is worth revisiting the definitions of several crucial terms. Hazard may be defined as a source or cause of potential harm or injury. Threat is an identified source of potential harm or injury, often implied to be manmade or intentional. Risk is the probability of harm or injury occurring. A vulnerability is an oversight or weakness that can be the method or avenue via which a hazard or threat causes actual damage or injury.

Risk is therefore a probability, a percentage, a likelihood. Risk also encompasses the exposure of people and property, and ultimately the result, impact, or cost of that exposure (Haddow, Bullock, & Coppola, 2008, p. 68). Assessing such risks is the bedrock of emergency management.

### Methodologies

There are multiple schools of thought regarding how to best assess risk for a given group, jurisdiction, or organization. Each has its own strengths and merits. In this context, the text will examine four such methodologies, each of which has the potential to be highly useful to an emergency manager establishing a baseline for his or her operation. Such assessments will be the basis for any holistic emergency planning effort, particularly in a complex urban environment.

#### *The Composite Exposure Indicator Method*

When attempting to gauge the level of risk for a given operation, data are absolutely vital. The composite exposure indicator (CEI) approach recognizes the importance of data, and builds upon it as a core function of a larger whole.

CEI assesses the risk from a given hazard based on a number of variables. It then produces a ranking based on selected hazards. The methodology essentially uses a mean and a standard deviation to predict the likelihood of a given hazard occurring for a given location. Federal Emergency Management Agency (FEMA) databases are utilized to provide the base information. Thus it can be easily visually referenced in chart form, essentially in a side-by-side comparison.

The 14 recognized variables utilized by the CEI method are:

- Hospitals
- Population
- Nuclear Power Plants
- Toxic Release Inventory
- Public Water Supplies
- Superfund Sites
- Sewage Treatment Sites
- Utility Lines
- Airports
- Roads
- Railroads
- Pipelines
- Dams
- Bridges

The variables are ultimately expressed as densities per square mile, thus providing a standardized comparison of all 3140 counties in the United States (including a handful of unique cases such as independent cities and parishes). The standard format also provides a minimum and maximum number of each variable occurring per county.

Using these data, several strong correlations begin to emerge. The number of hospitals and population are highly correlated, as are the number of public water supply systems and sewage treatment sites. Per federal risk analysis, the results indicate that the first and largest principal components are the transportation variables (road and rail) and the water-related variables (public water supply, sewage sites, dams, and bridges).

Larger CEI values imply that more people, critical facilities, and lifelines are exposed to potential damages from various natural and technological hazards. In other words, not surprisingly, emergencies tend to occur along key transportation and infrastructure corridors. Further simplified, urban areas have more infrastructure and people, and thus more risk.

This "big picture" tool is an excellent place for an emergency manager to start a more detailed risk assessment of his or her organization or jurisdiction. But its usefulness does not end there. The national-level database and assessment model can be utilized elsewhere, or on a varying scale, simply by altering the variables.

As will be discussed further, the CEI methodology (Haddow et al., 2008, p. 68–69), and generally all risk assessment, may be broken down into six key steps, including identifying and assessing the hazards, estimating risk and cost, and determining a course of action.

If an emergency manager for a town in Oregon wanted to assess earthquake risk, for example, he or she could start by plugging in seismic activity as a variable in the model. Clearly, his or her jurisdiction would have a higher risk than areas with no history of earthquake activity. Yet by utilizing comparative data from other sources such as the US Geological Survey (USGS) National assessments and their counterparts in other nations, the manager could build a very robust table based on these sources, including minimums and maximums as well as mean and standard deviation. A slightly deeper dive would then allow for correlations to begin emerging from the data.

In essence, the nationwide CEI tool utilized in the United States could be extrapolated to be useful in virtually any locale. As long as comparable data are fed into the model for a given, it will provide useful information in terms of overall risk. From an emergency-management standpoint, the CEI method is a very good place to begin.

### The CARVER Method

Another well-established type of risk assessment is the CARVER method. While the core CEI method relies on FEMA-supplied data modeling, CARVER is based upon US military physical threat assessments for facility security.

The CARVER acronym represents criticality, accessibility, recuperability, vulnerability, effect, and recognizability. CARVER was originally used in both an offensive and defensive assessment of a given location's vulnerability to assault. However, in the years since its original inception, it has been adopted by a broader range of experts, including the US State Department and many private-sector security companies and consultants. The methodology is easily translatable to emergency management as well.

In practice, CARVER essentially produces a matrix relating to a given asset or facility. Each component of the facility is assigned a numerical value, usually ranging from one 1 (low) to 5 (high). This simplicity allowed the method to be used quickly and in relatively sparse environments by US Special Forces during the Vietnam conflict.

One can envision a commando, his face smeared with camouflage paint, surveying a potential target from the undergrowth, noting vulnerabilities and other traits on a scrap of paper while planning an assault. One could also imagine a defensive position being thoroughly assessed to counter the same type of risks.

And though emergency managers face a less militarized environment on a daily basis, the same principals apply. While the CEI method tends to be broad and examine large geographic areas and general hazards, the CARVER matrix focuses down to a very detailed facility level, looking at specific threats in relation to specific subcomponents.

The key to an effective CARVER analysis is an understanding of how individual components work together and interrelate. Performing a simulated CARVER assessment of a hospital facility serves as an excellent example.

To effectively assemble an operational matrix, a number of critical components and systems must be examined. Electrical power and water supply are obvious places to start. Ventilation systems and logistical resupply systems would also be considered. Ability of staff and patients to access the facility would be key. Emergency systems such as generators and battery back ups would be worth considering.

It is here that CARVER's flexibility comes into play. The sample matrix here is relatively simple and takes a high-level look at critical systems in a general hazard condition. However, each system could be broken down into further subsystems for greater detail and understanding. For example, instead of listing "electrical supply," a detailed CARVER matrix could provide an overall score built upon scoring for multiple subsystems such as number of power feeds, reliability of grid, transformer condition, minimal required load, availability of qualified maintenance staff, and emergency power supply (Table 5.1).

Thus CARVER represents a scalable methodology for users. To cover a wide geographic area, an emergency manager might be best served by generating

**Table 5.1** A Sample CARVER Matrix for An Urban Water-Processing Plant

| Asset/System | C | A | R | V | E | R | Total Score |
|---|---|---|---|---|---|---|---|
| Chlorine tanks | 4 | 2 | 2 | 2 | 4 | 2 | 16 |
| Electrical substation | 3 | 3 | 2 | 3 | 2 | 5 | 18 |
| Pipe delivery grid | 3 | 4 | 3 | 3 | 3 | 2 | 18 |
| Pumping station | 3 | 3 | 3 | 4 | 4 | 3 | 20 |
| Purification tanks | 3 | 2 | 4 | 2 | 3 | 2 | 16 |
| SCADA system | 4 | 4 | 3 | 3 | 3 | 2 | 19 |
| Water intakes | 4 | 2 | 3 | 3 | 2 | 4 | 18 |

simple matrices for the largest, most overarching infrastructure systems. To closely assess a few high-value nodes of critical infrastructure, a small number of highly detailed and nuanced matrices may better fit the need. In either case, the scalability and modular nature of CARVER has the potential to prove very useful for emergency-management professionals.

### *The Risk Analysis Security Survey*

As CEI has a data-driven emergency-management background, and CARVER owes its conception to a military pedigree, the risk analysis security survey emerged from the private-security sector. It shares many traits with the methodologies already examined here, but there are important differences as well.

Broder cites three specific divisions of risk in his survey introduction—personal, property, and liability (2000, p. 1). The personal could also be interpreted as personnel, since it related to the human assets affected by risk. The property division focuses on tangible material assets such as buildings or vehicles. The liability component may be seen as legalities related to aspects of risk.

The security survey defines risk assessment as a rational and orderly approach, and a comprehensive solution, to problem identification and probability determination (Broder, 2000, p. 2). In this sense, it closely aligns with our prior definition of risk.

The assessment advocated by this methodology focuses on three key elements. Assets are what an organization or jurisdiction owns, designs, operates, or controls. Exposure is a list of specific hazards or threats that could cause damage, loss, or injury. Losses relates to the projected or calculated subtraction of assets which would result. These three categories may be seen as the first crucial steps in assembling a checklist that comprises the core of the security survey.

The survey and assessment then pivots to the internal mechanisms that can be controlled or influenced to reduce risk. These categories may be seen as the remainder of the survey checklist:

- Internal and external crime losses
- Logistics exposure
- Emergency and disaster planning
- Liability of officers or officials
- Environmental, health, and safety controls
- Damages to property by natural or human-caused hazards
- General liability for organization or jurisdiction
- Interruption of business or public services
- Other liabilities including professional, product, or insured

The private-sector origination of the security survey is certainly on display within this checklist. An important contrast it makes with the other methodologies examined here is its focus on organizational processes and procedures. Whereas the CEI, and to a lesser extent the CARVER, methodology focuses in large part on external threats or hazards, the security survey is much more introspective and policy-oriented. This trait makes it particularly useful within complex or layered organizations whose very actions may be exacerbating risk, or within operations with an identified internal threat component.

Finally, the security survey methodology proceeds beyond assessment and offers suggestions for process improvement. Once a specific risk has been identified, the security survey advocates for a three-tiered approach to minimizing or mitigating its effects:

- Implement established practices and procedures to minimize occurrence
- Design mitigation plans to minimize the loss if the incident does occur
- Develop detailed contingency and recovery plans to assure continuity

### *The FEMA Process*

The FEMA advocates a five-step process to a thorough risk assessment (2007, p. 64–68). Each step builds upon the one before to offer a holistic risk model for a given asset or system.

- Threat identification and rating: The threat is identified and defined. This step determines the types of threats or weapons against which the asset must be protected.
- Asset/consequence value: The facilities, functions, and capabilities that need to be protected are identified.
- Vulnerability assessment: The potential vulnerability of the asset against all identified threats is determined, taking into account all existing layers of security and redundancy.
- Risk assessment: The values for threat, asset, and vulnerability are multiplied to arrive at a risk value. This step will identify which threats or weapons pose the greatest concern.
- Mitigation options: Informed by the previous steps and resulting data, this stage provides potential actionable options to reduce or eliminate specific risks.

This process in some ways represents an evolution of the CARVER methodology in that it focuses on kinetic or active threats. Its *threat x asset x vulnerability = risk* model also mirrors the data-driven CEI methodology.

The FEMA method also offers several levels or tiers of detail, depending on the need.

- A Tier 1 assessment is sometimes referred to as a 70% solution and will involve a site review and a basic overview of facility engineering and operations.
- A Tier 2 assessment goes further, including a full on-site evaluation examining interdependencies, vulnerabilities, and mitigation options.
- A Tier 3 assessment is generally reserved for extremely high-risk assets and will include all previous efforts, plus complex, multithreat modeling, and structural analysis.

Taken as a whole, the FEMA methodology thus allows for a significant level of flexibility when assessing a given asset or facility.

### *Other Methodologies*

Regardless of what methodology or methodologies an emergency manager chooses to employ for his or her operation, the act of performing an effective risk assessment is absolutely critical. Despite the differences cited among the systems examined here, there are certain commonalities to all effective risk assessments. As noted, Haddow et al. (2008, p. 68–69) provide a nice summary of these six characteristics:

- Identify and characterize the hazard
- Evaluate each hazard for severity and frequency
- Estimate the risk
- Determine the potential societal and economic effects and any indirect costs
- Determine an acceptable level of risk
- Identify risk-reduction or mitigation opportunities

Many other systems and methodologies exist, some general in nature and some highly specialized. The Risk Matrix Approach is similar to CARVER in that it summarizes general hazards in a table format according to frequency and severity. The Threat and Hazard Identification and Risk Assessment (THIRA) is a system introduced by the US Department of Homeland Security (DHS) in recent years to assist jurisdictions in risk identification.

The Hazard United States (HAZUS) methodology has been in use for many years to help determine risk associated with earthquakes. And the Multiple Coastal Hazard methodology focuses on seven potential hazards to organizations or jurisdictions located along shorelines.

Active emergency managers and students in the field are strongly encouraged to more thoroughly review these methodologies, particularly if one of the more specialized systems may apply to their organization or community. No single system is perfect, and a consistent assessment methodology across multiple assets or systems may be seen as the true objective.

### The Product

Ultimately, any risk assessment is only as good as the resulting reporting. Each method examined here can generate useful and usable data. Yet no emergency manager exists in a vacuum. It will be crucial to translate these findings to a broader audience. In a public-sector role this may mean firefighters and police officers, or public-works managers and elected officials. In a private-sector role, this may mean business-continuity and technology experts, or executive officers of the company.

Whatever the assessment method or methods utilized, the results must be packaged in a presentable and useful format. This might be a packet led by an executive summary, and containing detailed charts and graphs displaying findings. Or it may be a digital presentation providing translatable and relatable visuals for the core findings of the assessment.

Incident reports, accurate records of unusual events in a standardized format, can provide excellent data points for an assessment (Somerson, 2009, p. 109). Such tangible and accessible data will be vital to identify frequency, trends, and to determine the effectiveness of risk-management programs and protocols.

Whatever the format, make it targeted to your respective audience and their needs. A thorough risk assessment is a complex and time-consuming process, and it is in the absolute best interest of the emergency manager and his or her organization that the results be put to the best possible use.

Finally, it is vital to note that a thorough risk and hazard assessment is not a one-time process or product. It is rather a cycle of analysis and application across an organization or jurisdiction. Effective reporting and data will strengthen the process throughout the cycle.

## EMERGENCY OPERATIONS PLANNING

While a thorough risk assessment may be thought of as the research arm of emergency management, operations planning is the practical translation and application of those findings. They are two sides of the same coin, and one is far less effective in the absence of the other.

Planning is a key component of a successful emergency-management program, yet it is often overlooked prior to an incident or disaster occurring. This can be especially true in a jurisdiction or organization that has not recently suffered a catastrophic threat or catastrophic loss. Note that the plan should be verified via drills and exercises once completed. (Information on the Planning Section of the incident management system may be found in Chapter 6, Incident Command Systems and NIMS.)

The professional security association ASIS International defines an emergency plan as a document outlining "the actions to be taken by an organization to protect employees, the public, and assets from threats created by natural and manmade hazards" (2003, p. 3). Such a document has a number of titles, varying slightly by organization. It may be called an emergency-management plan, an emergency response plan, or simply the emergency plan. This author prefers the title emergency operations plan (EOP), because it not only meets National Incident Management System (NIMS) standards but also connotes breadth and action.

Ultimately, the title of such a planning document matters far less than the content. An effective plan determines courses of action *prior* to an actual crisis situation. As discussed, performing a holistic risk assessment is a vital step. For planning purposes, it is also crucial to see what you have available. Previous plans, even ones that have become obsolete, may offer critical insights into the existing structure of the organization or jurisdiction.

It is important to set a defined scope and to understand the key components. What facilities or infrastructure must be accounted for by the plan? What personnel or visitors must be covered, and what are their demographics? Is this a one-person effort, or are you able to bring in other team or committee members to assist? Finally, gaining the support of upper management will be vital to the successful implementation of any comprehensive EOP (ASIS, 2003, p. 5).

## Plan Components

All of these efforts essentially constitute planning to plan. With the right pieces in place, it is time to craft the document itself. NIMS provides some excellent guidance on what must be included, and the format is appropriate not just for American jurisdictions but for nonprofit organizations, corporations, and for any governmental agency in a developed nation.

The planning document should, unsurprisingly, begin with an Introduction. This portion should identify the organization, and state the need or motivation for developing an EOP. A strong purpose statement providing justification for the plan is an excellent component of the introduction. The scope and methodology should also receive at least brief mention here, noting what research has been done and how the plan was compiled and assembled. These will assist the reader in understanding how the plan was assembled.

The next major section is the Situation. This segment contains some of the hard data compiled during the research phase, and sets the tone for the core of the document. It should include the findings of the risk and hazard analysis/assessment. It should also give an overview of the organization or jurisdiction, and present key assumptions made during the planning process. It may also include a broad listing of emergency or disaster categories, to set the stage for more detailed discussions later in the document.

The Roles and Responsibilities segment comes next. This portion details, one by one, the organizations or institutions that will be involved in any emergency or disaster impacting the assets covered by the plan. It should start with the responsibilities of the organization or jurisdiction itself, and then expand outward. Individual sections may include local government, state or regional government, federal government, nongovernmental or volunteer organizations, and professional organizations. Existing memoranda of understanding may be included here or in a supporting appendix (Fig. 5.1).

The core of the document follows. Referred to as the Concept of Operations or ConOps, this is where the plan shifts from an overview to the implementation level. It begins with a General Status segment, reviewing standard

**FIGURE 5.1**
This 2016 construction crane accident in New York City activated the city's structural collapse plan.
*Credit: Shutterstock.*

operational procedures and emergency operational procedures. This is also a good point to introduce the concept of coordination and control, which will again note every agency or department that will be involved in an emergency response, and define how they will interact during a crisis. Segments on information gathering and analysis may be useful, as will reviewing communications and notification procedures.

The Concept of Operations section then moves on to the individual phases of a disaster or crisis: prevention, preparedness, response, recovery, and mitigation. Operational guidelines and expectations for each phase should be clearly defined. Parameters should be established for each phase so that individual units may clearly identify within which stage or stages of a disaster they are operating.

The next segment relates to Finance and Administration. This portion should clearly differentiate standard administrative tasks and practices from those which take priority in an emergency. Well-defined financial-management practices and responsibilities should also be outlined.

A section on Continuity of Operations (COOP) follows. This segment should provide a clearly defined command structure during crisis, and may even include a NIMS-compliant organizational chart. It will also give lines of succession identified by position or role should any key figure be unable to fulfill his or her duties due to injury, travel, etc. COOP details should also list possible temporary facilities, including spaces for and emergency operations center (EOC), backup data centers, and other key functions.

A series of Functional Annexes conclude the main body of the EOP. These will focus on the specific threats identified by the hazard analysis and assessment at the beginning of the process. These briefs are organization-specific and may include fire, earthquake, active shooter, civil unrest, etc. Each annex should be kept extremely short and concise, no more than one or two pages in length. Individual annexes function as bullet points or checklists for critical actions mandated by a specific hazard. Separate annexes may also be included detailing individual mitigation or response measures such as evacuation or shelter-in-place.

A series of appendices and references are added to the end of the base plan. These documents provide additional information and context to the whole, and may include EOC protocols, maps or floorplans, and emergency contact lists.

A sample table of contents for an EOP follows, providing additional detail.

## SAMPLE EMERGENCY OPERATIONS PLAN (EOP)

**XYZ Corporation**

*Table of Contents*

By following this basic outline, an emergency manager can craft a realistic and valuable tool to prepare for future crises. Whenever possible, the senior elected official or chief corporate officer should provide a letter of promulgation, authorizing and endorsing the emergency plan under his or her authority. This step is critical not only to alert executive management to the status of the program but also to ensure buy in and support from other managers. Version control will also be vital, and a simple form tracking any changes or edits to the core text should also be included.

Finally, as an emergency manager there is absolutely no need to go it alone. In addition to a wealth of publications and online resources related to emergency planning, any large organization or urban jurisdiction is going to have a number of useful subject-matter experts. Do not hesitate to reach out to experts in fields such as technology, transportation, law enforcement, fire services, urban planning, and emergency medicine. In many ways, the modern emergency manager is expected to be a generalist with the ability to liaise with a wide range of professionals. An effective EOP will ultimately have one primary author, but many contributors.

## CASE STUDY: THE CHEMICAL, BIOLOGICAL, RADIOLOGICAL THREAT

In the past, terrorist organizations and other extremists were unlikely to initiate chemical or biological attacks. Consider the late 1970s as a reference point. At that time, the two most visible terrorist organizations were arguably the Irish Republican Army (IRA) and various factions of the Palestinian Liberation Organization (PLO). Both were centered upon territorial or nationalistic goals, and thus had a core goal of swaying public opinion—winning "hearts and minds."

Hoffman, among many other authors and experts, notes that this is simply no longer the case. He cites the 1995 sarin gas attack in the Tokyo subway, the 1995 bombing of the federal building in Oklahoma City, and the 1998 bombings of two American embassies in Africa (2006, p. 270–271). These turning points are very noteworthy because the four attacks took place within a very short time span, and were perpetuated by an Asian religious cult, a disgruntled former member of the US military, and an Islamic extremist group. It would be difficult to find anything in common with the three diverse perpetrators, except for the fact that they were willing to use weapons of mass destruction (WMD) to cause massive casualties. Somewhere between 1975 and 1995, the principles and goals of such organizations had morphed into a range of extremist groups more than willing to run up the civilian body count in the name of their particular cause.

If there was any remaining doubt about the shift in this paradigm, it was erased when the same Islamic extremist organization that had bombed the embassies in 1998—Al Qaeda—utilized airliners packed with civilians as cruise missiles on the morning of September 11, 2001. Unlike the PLO in the late 1970s, such organizations now view civilian passengers not as hostages or bargaining chips but merely as collateral damage.

In terms of the knowledge necessary to utilize chemicals and biological agents as weapons, the Weapons of Mass Destruction (WMD) Student Manual published by the DHS offers some important insight. The text notes that "terrorist organizations have greater access to WMD materials and the technical expertise to weaponize them" (Center for Domestic Preparedness, 2007, p. 5–6). It goes on to note the factors that may make an attack more damaging or lethal, including surprise, means of attack, target of attack, and foreknowledge of response techniques.

In fact, Hoffman indicates that Al Qaeda has come to the same conclusion, noting that "incontrovertible information has repeatedly come to light that clearly illuminates Al Qaeda's long-standing and concentrated efforts to develop a diverse array of chemical, biological, and even nuclear" weapons systems (2006, p. 272). The same could be said of multiple other extremist groups, and thus the risk largely shifts to determining which types of materials could be operationalized.

It is important to note that a number of chemical agents have been occasionally utilized by military organizations since World War I. These include irritants such as tear gas but also several potentially lethal classes of substances (Bevelacqua & Stilp, 2009, p. 12). The three primary categories are neurotoxins (nerve agents), asphyxiants (blood agents), and respiratory irritants (choking agents). In sufficient quantities, multiple agents within each of these categories could prove fatal, including chlorine, mustard gas, and sarin. It is important to note that the effects of chemical weapons are seen very soon after use and could be referred to as highly acute.

Biological agents were also explored by multiple militaries, including the United States and the Soviet Union in the midst of the Cold War. However, their development and deployment seems to have lagged significantly behind that of their chemical cousins. Broadly categorized as bacteria, viruses, and toxins, potential biological agents include anthrax, smallpox, and ricin. It is also important to note that the effects of many biological agents may not appear for weeks or even months.

The radiological threat, or so-called dirty bomb, is also a clear and present danger. The former Soviet republic of Georgia serves as a prime example. Its security services report interdicting at least three attempts—in 2001, 2006, and 2016—at smuggling highly enriched radiological material through its territory. If such material was placed around a traditional explosive device, it would disperse radiation over the width of the blast radius as well as downwind from the explosion. And while such a device would not be a direct or significant threat to public health outside of a limited geographic area, it would cause widespread panic as the public would likely have difficulty differentiating between a nuclear and a radiological explosive event.

Recognizing these developments, a holistic risk assessment must conclude that the United States and its Western allies are increasingly likely to face a chemical, biological, or radiological attack. A real-world example of an extremist group working with chemical and biological agents displays the type of threat vector urban emergency managers are most likely to face.

Kortepeter and Parker note that the Japanese cult Aum Shinrikyo not only executed a widely reported 1995 sarin gas release in a Tokyo subway but also had a well-funded and extensive effort to develop biological weapons including anthrax and botulinum (1999, p. 523). Despite years of effort and significant funding, Aum was unable to weaponize the biological agents into anything truly operationally deployable.

Instead, the Aum followers punctured plastic bags on crowded train cars containing the chemical sarin in an attack that killed 12, critically injured 50, and caused respiratory distress for some 5000 others. The United States has considerable experience in responding to and recovering from hazardous materials incidents, and such responses would more or less correspond to a chemical attack (Sylves, 2008, p. 118). Response to a biological release would be more dependent on the public health and epidemiology functions. Unlike the acute and immediate threat posed by chemicals, most biological agents have a prescribed course of medication and treatment capable of sparing at least some victims. The radiological threat is one that may be effectively countered on a tactical level utilizing the containment policies and procedures practiced and implemented at nuclear power plants, yet would cause more widespread public alarm.

Likelihood may also rest on simple availability. Commonly used industrial materials such as chlorine and ammonia have the potential to be utilized during a terrorist act. They regularly travel our nation's highways and railways in giant tankers. Biological agents such as smallpox and anthrax are simply not readily available in any such quantities. Oftentimes, accessibility is a final and determining factor in an extremist group's operational decision making.

In many ways, the three threats are simply a faster (chemical), slower (biological), and more complicated (radiological) version of the same intended impact. Ultimately, given the wide range of weapons that terrorists and extremists have used in the past, it must be presumed that WMD is a genuine risk to Western cities, and thus a threat for which emergency managers must be prepared.

## SUMMARY

Two vital tools will make or break the efforts of the urban emergency manager. The first is a risk analysis or assessment. There are a number of differing methodologies for performing this crucial task, and emergency managers should select the best type or types to fit the specific needs of their organization. Any holistic emergency plan will be based upon this assessment.

Next, the emergency manager must draft an emergency planning document for the organization. The details of a NIMS-complaint plan are outlined here, and a sample table of contents for such a plan provides additional detail and context. The two tools will serve an emergency manager well in an actual crisis, and other subject-matter experts should be consulted to produce the best assessment and plan possible. A wide range of threats, including chemical and biological agents, must be accounted for by such plans.

## References


ASIS International (2003). *Emergency planning handbook*. Arlington, VA: ASIS.

Bevelacqua, A., & Stilp, R. (2009). *Terrorism handbook for operational responders*. Clifton Park, NY: Delmar.

Broder, J. (2000). *Risk analysis and the security survey*. Burlington, MA: Butterworth-Heinemann.

Center for Domestic Preparedness (2007). *WMD student manual*. Washington, D.C: U.S. Federal Government.

Federal Emergency Management Agency (2007). *Site and urban design for security*. Washington, D.C.: U.S. Federal Government.

Haddow, G., Bullock, J., & Coppola, D. (2008). *Introduction to emergency management* (Third edition). Burlington, MA: Butterworth-Heinemann.

Hoffman, B. (2006). *Inside terrorism*. New York: Columbia University Press.

Kortepeter, M. G., & Parker, G. W. (1999). Potential biological weapons threats. *Emerging Infectious Diseases*, *5*(4), 523–527.

Somerson, I. S. (2009). *The art and science of security risk assessment*. Arlington, VA: ASIS.

Sylves, R. (2008). *Disaster policy & politics: Emergency management and homeland security*. Washington, D.C: CQ Press.


## Further Reading


Federal Emergency Management Agency. *Risk assessment approaches*. Available from: <https://www.fema.gov/media-library-data/20130726-1545-20490-4252/mhira_ra.pdf>. Retrieved on June 2016.

National Commission on Terrorist Attacks Upon the United States. (2002). Washington, D.C.: U.S. Federal Government.

CHAPTER 6

# Incident Command Systems and NIMS

## INTRODUCTION

The summer of 1970 was a particularly hot and dry one for central and southern California. Winds were relentless and temperatures reached the 100°F mark on several occasions. The first true wave of urban sprawl had come to the state in the previous decades, resulting in subdivisions and shopping malls cropping up in what would be considered rural backcountry just a few years prior. The US Forest Service still had the true responsibility for safety over many sparsely populated areas, supported by a patchwork of local and state firefighters, many of them volunteers. The current mandates called for firefighting resources to be operated under a statewide disaster plan, which represented more of a loose confederation than a binding document.

In late September, in the high country near Oakland, the inevitable finally occurred and a large and aggressive wildfire was born. Within days, there were literally hundreds of fires burning within the state. During a 2-week period, more than 750 individual wildfires would sweep across southern and central California, burning more than 500,000 acres. By the time the last major wildfire was reported to be contained on October 4, 16 lives had been lost, along with more than 700 homes.

The vast majority of the damage occurred at the edges of the new urban and suburban sprawl. Firefighters refer to this zone as the "interface"—the space where human development and wild backcountry blur together. Nearly 20,000 firefighters had waged a fierce and tireless battle against the optimal fire conditions and challenging terrain. Despite their efforts, all involved recognized that the operations had not gone smoothly. After-action reports at the agency, county, and state levels all cited the same concerns: communications, planning, intelligence, and logistics.

No one implied that the firefighters had not worked hard to contain the multipronged disaster; the finger of blame was instead uniformly pointed



Urban Emergency Management. DOI: http://dx.doi.org/10.1016/B978-0-12-805307-2.00006-8

at a lack of organization. And so as exhausted firefighters finally got some sleep, and communities began to recover and rebuild, all involved knew something had to change.

## ANALYSIS

### A Little History

For millennia, humans have banded together during times of crisis, battling each other as often as they battled nature. Command and control structures tended to make leaps forward during such periods of armed conflict or similar adversity, but varied widely across regional and cultural divides.

Leap forward to the late 1960s and the early 1970s, as a number of devastating wildfires claimed an alarming number of lives and acreage across the western United States. The public outcry following the devastating California fires in autumn of 1970 forced Congress to act, mandating that the US Forestry Service come up with a more durable solution. And firefighters in California once again took the lead.

In 1973, the first FIRESCOPE technical team was established to guide this research and development. (The FIRESCOPE acronym represents a rather forced FIrefighting REsources of Southern California Organized for Potential Emergencies.) According to the Federal Emergency Management Agency (FEMA), a defined incident command system (ICS) was one of the most important concepts to emerge from this work. Rather than trying to piece together an effective command and control format assembled from various agencies with their own variants, ICS offered the promise of a modular yet unified command model.

The wildland firefighting culture itself was also a key source of this systemic development. Hillyard notes a firm group expectation for both a top-down and bottom-up willingness to innovate. He notes that frontline practitioners were expected to be "trained professionals who are oftentimes the impetus for new ideas, concepts, and innovations" (2000, p. 159). In turn, the organization's senior leadership consisted almost exclusively of individuals who had previously worked a variety of frontline positions, making them more apt to understand tactical issues and to recognize emerging talent.

Widely adopted by the wildland fire community, ICS then soldiered on in relative anonymity for decades with minimal recognition from outside agencies. The early 1990s again saw a number of large fires in the western United States as well as the bombing of the World Trade Center in New York and of the Murrah Federal Building in Oklahoma City. The large-scale responses by multiple departments across multiple levels of government again changed the playing field.

In the late 1990s, following multiple studies and analyses of such incidents, a five-phase model of crisis management emerged. Roughly aligned with the five phases of emergency management cited throughout this text, scholars Pearson, Misra, Clair, and Mitroff (1997) sought to define the sequence of events experienced by any network or organization during a crisis. Those phases are:

- Signal detection
- Preparation
- Damage containment
- Recovery
- Learning

During this period, a broader recognition of the value of crisis management—and of ICS in particular—began to truly take hold and wider scale adoption took place, particularly among highly specialized or regional units. Yet it would take an even larger shock to the system to drive full implementation.

## National Incident Management System Emerges

Following the September 11, 2001, terrorist attacks in the continental United States, a new agency was created to combine a number of existing departments and functions under one chain of command. The Department of Homeland Security (DHS) instantly became the second largest department in the US federal government, trailing only the long-established Department of Defense (DoD).

As was necessary and to be expected, there was a fair amount of soul searching after the deaths of nearly 3000 American civilians. The National Commission on Terrorist Attacks Upon the United States, better known as the 9/11 Commission, became the most visible advisory body related to the attacks. Extensive guidance and recommendations were provided both in a voluminous public report and via classified missives within the federal government.

One outcome of this process was a series of statements or declarations from the office of then-President George W. Bush. Known as Homeland Security Presidential Directives—or HSPDs—these documents became crucial to the overall domestic-security posture of the United States. Furthermore, they also mandated significant changes within the emergency-management function at both the local and state levels. (See Chapter 8, Emergency Management and Homeland Security in Urban Areas, for further context.)

One of the most important of these documents was HSPD-5, issued on February 28, 2003. Much of the document is simply updating or "tweaking" previous HSPDs. However, in its Statement of Purpose on page one, HSPD-5 specifically mentions the goal of managing "domestic incidents by establishing a single, comprehensive national incident management system." It is important to note that at this time, the federal government was still in the midst of combining a multitude of existing agencies and departments into the DHS.

As noted by Haddow, Bullock, and Coppola, "the creation of DHS was the culmination of an evolutionary legislative process that began largely in response to criticism that increased federal intelligence interagency cooperation could have prevented the September 11 terrorist attacks" (2008, p. 14). The aforementioned internal investigations within other federal agencies were reaching similar conclusions, and despite more than one agency's complaints about losing administrative independence, the DHS was here to stay.

HSPD-5 is interesting in that it placed much greater focus on the vertical integration among various levels of government than on the horizontal organization and alignment that was the focus of most other HSPD documents. Unquestionably, the most critical element of HSPD-5 was the comprehensive incident management system mentioned in the document's Purpose statement.

A number of other important statements are included in HSPD-5—including the assignment of the Attorney General as being ultimately responsible for criminal terrorism-related investigations within the United States. But none are more central and far-reaching in terms of emergency response than is the creation of the National Incident Management System (NIMS) included in the Tasking portion of the document. This section of HSPD-5 is where the bureaucratic reorganization is translated into actual practices and procedures in the field—where the rubber truly hits the road. It designates that all federal departments will align their organizations to the task of national-security coordination. It further outlines the creation of the National Response Plan (NRP)—later the National Response Framework (NRF)—which establishes the way forward.

Yet even in this early document, NIMS was clearly central to the training and response elements related to any homeland security incident or situation. It required "governments to work effectively and efficiently together to prepare for, respond to, and recover from domestic incidents, regardless of cause, size, or complexity" (2008, p. 4). This may be the single most-important phrase in the entire HSPD series, as it requires governmental agencies at *all* levels to be prepared for *all* hazards. It furthermore requires a unified system of incident management that can be utilized by each and every responding agency. This represents nothing less than a complete paradigm shift in the field of emergency management, which had primarily utilized a disparate collection of localized systems to respond to crises and disasters.

The resulting atmosphere of (mandatory) cooperation across jurisdictional lines has been extraordinary in its impact. The amount of cooperation in training, exercises, and drills has greatly increased as a direct result of the directive and of the initiatives which followed. In addition, NIMS has become the operating system to which all emergency-response agencies—regardless of size or location—are expected to comply. In the United States, it is not merely big-city governments employing the concepts of ICS and NIMS. Rural sheriff's departments utilize NIMS incident command. State emergency-management agencies utilize NIMS incident command. DHS utilizes NIMS incident command. Its simplicity and compartmentalization allow for an expandable or contractible command structure that is truly efficient in all-hazards situations, representing true vertical and horizontal integration. This is the most significant and lasting legacy of the HSPD series.

NIMS has therefore become one of the pillars of emergency management within the United States. From its inception, it has sought to standardize and streamline disaster response across a wide range of threats and vulnerabilities.

As described in the introduction to the 2008 version, NIMS provides "A systematic, proactive approach to guide departments and agencies at all levels of government, nongovernmental organizations, and the private sector to work seamlessly to prevent, protect against, respond to, recover from, and mitigate the effects of incidents, regardless of cause, size, location or complexity." This national system strives to represent a standardized and flexible command structure that incorporates the best practices of participating agencies. And, from this description alone, that would appear to be quite a daunting task.

It was further refined via the implementation of guidance documents such as the NRP/NRF, with which it is designed to be complimentary. The system was also designed to be capable of dealing with disasters of manmade and/or natural origin—clearly an all-hazards approach (Haddow, 2008, p. 119).

As stated in the Preface of the December 2008 version of the guiding document, "NIMS represents a core set of doctrines, concepts, principles, terminology, and organizational processes that enables effective, efficient, and collaborative incident management" (2008, p. 3). Thus, NIMS is not a "nuts and bolts" operational plan, but a systemic guide to disaster response.

From the first pages of the NIMS Introduction and Overview section, subsequent versions note the necessity for a multiple-hazards approach, specifically noting the terrorist attacks of September 2001 and the destructive hurricane seasons of 2004 and 2005. By clearly citing NIMS as a comprehensive, multijurisdictional approach to such a broad range of threats, the guidance document decisively stakes out an all-hazards viewpoint. This section goes on to emphasize that any successful management system must also be flexible and standardized. Finally, the Introduction and Overview addresses the individual components of NIMS, including: preparedness, communications and information management, resource management, command and management, and ongoing management and maintenance.

The segment directly addressing Command and Management is also crucial. In the very first paragraph, it defines incident management as including "directing specific incident operations; acquiring, coordinating, and delivering resources to incident sites; and sharing information about the incident with the public" (2008, p. 45). The ICS is then analyzed, breaking out into five functional areas: Command, Operations, Planning, Logistics, and Finance/Administration.

The core document then goes on to address the crucial aspect of successful management characteristics, breaking ICS into 14 proven operational management traits:

- Common Terminology
- Modular Organization
- Management by Objectives
- Reliance on an Incident Action Plan (IAP)
- Manageable Span of Control
- Predesignated Incident Mobilization Center Locations and Facilities
- Comprehensive Resource Management
- Integrated Communications
- Establishment and Transfer of Command
- Chain of Command and Unity of Command
- Unified Command
- Accountability of Resources and Personnel
- Deployment
- Information and Intelligence Management

The NIMS document also compares and contrasts the command functions in the form of a single Incident Commander (IC) versus Unified Command. Very importantly, specific functions of the Command Staff [(Public Information Officer (PIO), Safety Officer (SO), Liaison Officer (LNO)] and the General Staff [Operations Section Chief (OSC), Planning Section Chief (PSC), Logistics Section Chief (LSC), and Finance and Administration Section Chief (FASC)] are defined. NIMS also goes into some depth regarding the position-specific responsibilities that are assigned within each section, which will be reviewed later in the chapter.

Perhaps due to its flexible organizational structure, and its use by multiple public-sector agencies, NIMS has been adopted by a number of private-sector entities as well. Many hospitals and medical facilities employee some variation of the NIMS model, as do transportation companies and airlines. The nonprofit sector has also largely embraced the incident command concept, particularly in points of real-world interaction with public agencies. (Such collaborations are addressed in greater detail in Chapter 7, Collaboration in Urban Environments.) For example, the author clearly recalls pleasant surprise upon deploying to a hospital suffering a serious power failure during a bitterly cold Great Lakes winter, and finding an internal ICS staffing structure already up and running.

In a very big-picture sense in the United States, DHS has utilized the same model for protecting critical infrastructure, assigning sector-specific agencies (SSAs) to develop sector-specific plans (SSPs). For example, the Department of the Treasury is responsible for the banking and financial sector, the Department of the Interior is responsible for national monuments, and so on (Purpura, 2007, p. 361). The National Strategy for Homeland Security forms the framework of this overarching system, and is, in turn, NIMS based.

From managing by objective to a multiagency command structure, the difficult lessons learned in the West Coast wildfires of the 1970s, in the New York City response of 2001, and along the Gulf Coast in 2004 and 2005 are truly being implemented each and every day half-a-continent away. NIMS may not be perfect—and it does seem to be ever evolving—but it is far superior to what existed prior.

## NIMS Roles in Operational Practice

As any frontline manager can tell you, having a great concept on paper does not mean it is easily translated into practice. Many a detailed plan has been placed in a binder on a shelf and left to gather dust. Given the complexity and confusion surrounding any large-scale emergency, the concepts noted above faced a very real possibility of fading into obscurity.

On the contrary, NIMS has now become the de facto playbook by which all emergency responders operate on "game day." Not only have urban emergency managers, firefighters, and law enforcement adopted the language and the structure but also so have a vast number of nonprofit and nongovernmental organizations, hospitals, colleges and universities, and corporations. As Erickson notes, ICS "provide a flexible framework of management, rather than a definitive algorithm" (2006, p. 236). All of which leads to a vital question worthy of review: What does NIMS look like in the field? The best place to start is at the top, with the Command and General Staff positions (Fig. 6.1).

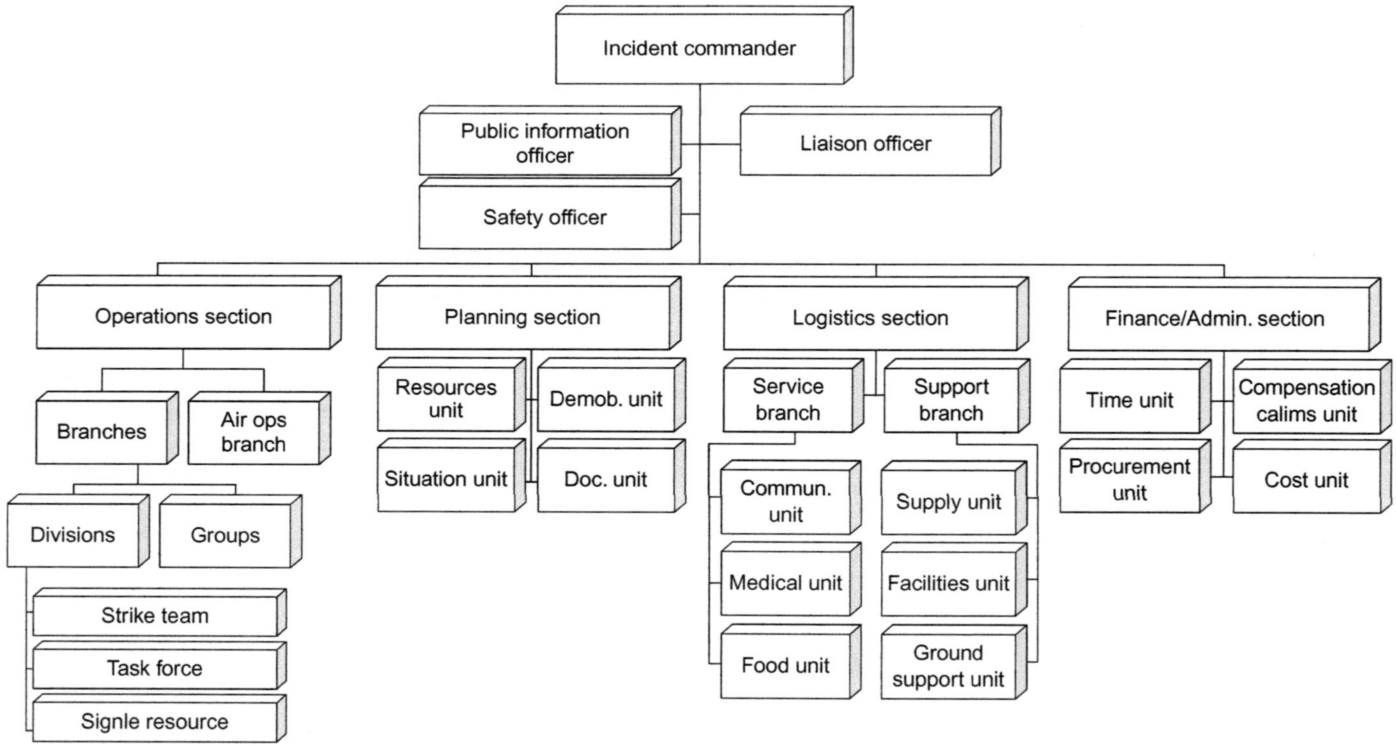


**FIGURE 6.1**
A sample NIMS organizational chart showing command and general staff, including the four major response sections. *Source: FEMA.*

### *Incident Commander*

At the top of the NIMS organizational response chart is the position of IC. As the man or woman with the overall responsibility for a given incident, there are few roles in modern society to which the maxim "the buck stops here" so aptly applies. Ultimately, he or she will be tasked with making life-or-death decisions for victims of the disaster and for responders in the field. The IC is unquestionably the captain of the ship. As a result, the IC is almost certainly an operational veteran, generally having served in multiple prior senior leadership roles across multiple deployments. A complete and thorough understanding of the ICS structure and function is vital.

Like all Command and General Staff roles, the IC should be qualified via the responsibility, authority, and expertise (RAE) principles. This means that the IC may not be the highest ranking individual on a standard day at the office—but a day in which NIMS is activated is not standard. Rather his or her qualifications grant the privilege and responsibility to take command of an active disaster scene. With an organizational chart guaranteed to be filled with truly Type-A personalities, this is easier said than done.

### *Command Staff*

Reporting directly to the IC are a handful of specialized professionals that make up the Command Staff. These roles consist of the Public Information Officer (PIO), the Safety Officer (SO), and the Liaison Officer (LNO).

The PIO is responsible for handling all nonoperational communications releases. He or she will interact closely with the media, will coordinate messaging with counterparts from other agencies or jurisdictions, and will function as the spokesperson for the Command and General Staff. The PIO is also responsible for mass messaging and mass notification during a large-scale incident, a function that may prove particularly crucial under rapidly changing or evolving conditions.

The SO is another key position within the Command Staff. As the name implies, his or her ultimate responsibility is the safety of the disaster scene. This wide-ranging role can involve everything from ergonomics at a command post to safe rappelling practices in a frontline search-and-rescue operation. Far from a "paper tiger," the SO is empowered to immediately halt all response and recovery activity if he or she witnesses an ongoing unsafe practice—an authority shared only by the overall IC.

The LNO is the final component of the Command Staff team. He or she will be responsible for productive and successful interactions with all personnel not within the IC's direct chain of command. This translates into professional interactions with local and regional elected officials, nongovernmental organizations, and unassigned public-safety personnel with a wide range of departmental backgrounds and expertise. A truly effective LNO cannot only smooth ruffled feathers and bruised egos but gain meaningful contributions to the emergency-response effort from a range of supporting organizations.

It is also worthy of note that, at the IC's discretion, an Intelligence Officer role can be added to either the Command Staff or the General Staff. This role may comprise intelligence in the traditional law enforcement or counter-terrorism sense, or it may serve an information-gathering function in a natural or unintentional human-caused disaster.

### *General Staff*

To make a comparison to the Executive Branch of federal government, if the Command Staff represents the President's cabinet of closest advisors, the General Staff represents the active department heads. From a more traditional organizational chart alignment, General Staff represent the managers who are accomplishing specific tasks and objectives related to the response effort. General Staff roles are referred to as Section Chiefs, who oversee Planning, Operations, Logistics, and Finance and Administration.

The PSC or Plans Chief is in charge of predicting stuff; responsible for looking in the crystal ball. That is to say, he or she is always looking two or three operational periods into the future to effectively gauge progress, needs, and requirements at some date down the road. This will require interacting with the other Section Chiefs to determine potential outcomes and shortcomings within their respective units. Such a predictive task is often very challenging in an urban response due to the complex level of infrastructure and high civilian population which may be affected (Fig. 6.2).

The OSC or Ops Chief is in charge of doing stuff; the one who has boots on the ground in the hot zone. Operations is responsible for virtually every dangerous or dirty job during a large-scale response ranging from firefighting, to search and rescue, to perimeter security. Balancing responder safety with victims' needs presents major span-of-control challenges during any large-scale incident. Given the scope and scale of an urban response, this may well be the most difficult role in the entire organizational chart.

The LSC or Logs Chief is in charge of getting stuff; oversees the unenviable task of supplying a massive response or recovery operation with everything it needs, from food and fuel to sandbags and soap. He or she must work closely with each and every Section Chief as an overabundance or lack of supplies will greatly impact the overall effort. In urban responses, commodity shortages, supply costs, and damage to transportation infrastructure often represent three very large hurdles for the Logs Chief.

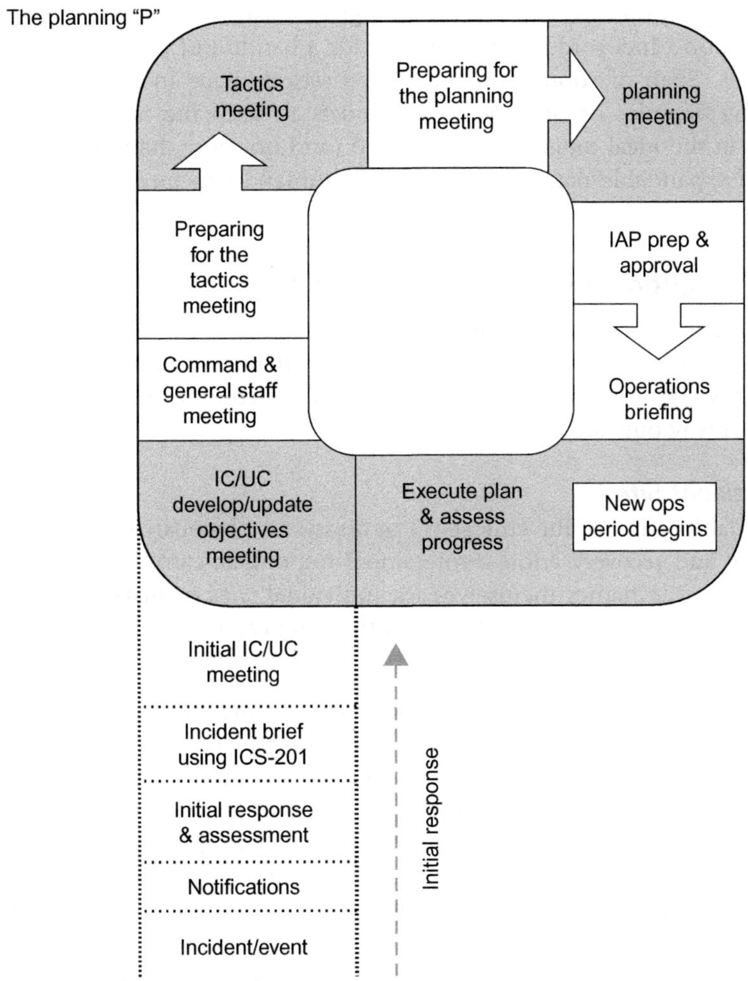


**FIGURE 6.2**

The "Planning P" makes an excellent visual reference for the complex role of the Planning Section within the response and recovery phases of a large-scale incident. *Source: FEMA.*

Finally, the FASC or Admin Chief is in charge of accounting for stuff; responsible for making everything else run. A thankless job, the role tracks expenses and performs administrative tasks as prioritized by the IC. Such a function is crucial in recapturing federal funding following a large-scale incident and allows other sections to focus on their immediate tasks and objectives rather than filling out forms and spreadsheets.

Depending on the size and severity of a given incident or series of incidents, each Section Chief could be responsible for a handful of professionals, or of hundreds. Span of control can become a serious issue in larger responses, requiring deputies or other subcommanders to keep the number of direct reports in the ideal range of five to seven (and no more than 10). The flexible and expandable nature of the NIMS command structure makes this task possible.

## NIMS Positioning in Operational Practice

Following this review of the major leadership roles within the ICS, it is crucial to examine the geographies of a major urban response. For a large or complex scene, it is most helpful to begin at the core of the emergency and work our way outward.

### *The Incident Site*

The actual location of the emergency or disaster is obviously the focus of any response and recovery effort. Professional responders carry on a long tradition of creating names themselves for individual sites or instances—such as Ground Zero or the Blue Cut Fire—yet there are certain key terms that apply to any large-scale incident.

The term hot zone refers to the absolute core of the emergency. It is where frontline response and recovery work is taking place, and is likely to be extremely dangerous. For this reason, an inner perimeter is often established surrounding the area to keep nonresponders out. Depending on the causation of the incident, the hot zone may include active aggressors, uncontrolled fires or hazardous materials leaks, potential structural collapse, or even radioactive materials. No personnel are authorized to enter the hot zone without the explicit direction of the OSC and thus the IC. Working time within this area may be strictly limited due to dangerous conditions, such as excessive heat or the presence of hazardous materials.

The warm zone is the geographic area immediately surrounding the hot zone. It contains some risk, and civilians will be kept at a distance. An outer perimeter may be established around the warm zone for this reason. The warm zone may include territory "reclaimed" from the hazard via the neutralization of active aggressors, the extinguishing of fires, or the containment of hazardous materials. Responders often use staging locations within the warm zone to ready tools or machinery, or to apply or remove personal protective equipment (PPE) as they transition into and out of the hot zone.

The cool zone is an area somewhat removed from the hot zone, where no immediate threats or hazards are present. This is not to say that it is completely free from risk. Once access control has been established at the incident site, civilians, the media, and the emergency operations center (EOC) will be located within the cool zone.

### *Forward Command*

The incident command post (ICP), sometimes called the forward command post, will be located as close to the actual disaster scene as is possible, often within the warm zone. A senior commander will be in charge of the tactical response and recovery efforts at this location—likely the IC, Deputy Incident Commander, or OSC. During the initial response phase, the ICP may be three or four individuals from different departments standing around a map spread across the hood of a truck. In an extended recovery operation, it may evolve into a compound of fully equipped, climate-controlled command trailers (Fig. 6.3).

The incident base, which serves as the forward command point for the Logistics Section, may be collocated with the ICP, as may be the staging area where personnel and equipment prepare to be activated. In an urban environment, ease of access and availability of space within the warm zone may limit how many individuals can be forward-deployed to this location.

### *Emergency Operations Center*

The EOC is a core component of the command and control function at a large-scale or complex emergency. Slightly removed from the frontlines, it essentially provides strategic guidance and high-level support to the entire Command and General Staff. An EOC manager may be designated to run the actual facility, yet

**FIGURE 6.3**
An example of a forward-positioned incident command post, in this case established for a training exercise in 2008. *Credit: Author.*

a senior staff member will always be present in person—often the IC. Multiple Section Chiefs will also run their operations out of the EOC.

The facility itself may be lavish or Spartan. However, it will be responsible for maintaining situational awareness of the incident scene as well as outside forces that could impact the operational situation (civil unrest, severe weather, etc.). As a result, the EOC is likely to be technology-heavy, with several computer systems, radios, satellite/cellular phones, and multiple projectors or flat-screen monitors. Emergency managers should be decisive and demanding about potential locations, which should be predesignated whenever possible.

Two documents will be key to the effective functioning of the EOC and its personnel: the emergency operations plan (EOP) and the incident action plan (IAP). As discussed in Chapter 5, Assessing Risk in Urban Areas, the EOP serves as the strategic or structural guiding document for an organization or jurisdiction. It provides high-level guidance for emergencies and recommends individuals/titles to be assigned to the EOC on the organization's behalf. It also sets overarching expectations for a response or recovery effort.

The IAP is a more tactical and practical document and serves as the Operations Section's primary guidance for each and every operational period. Assembled primarily by the Planning Section, the IAP examines a full operational period—usually 12–24 hours in length—and assigns individual task forces, strike teams, and units based on operational need. A good IAP also sets tangible and achievable objectives, tracks the current threat picture, weather, and important mission milestones. Frontline bosses view the IAP as nothing short of a religious document, and are loathe to go into the field without the most current copy.

### Unified and Area Command

The concepts of Unified Command and Area Command are crucial to complex incidents, or to multiple simultaneous incidents. Unified Command is a common tenant of any incident involving more than one responding agency or jurisdiction. It is, quite simply, an acknowledgment that the existing command structure may not be adequate to handle a complex or evolving threat that crosses recognized boundaries. As Hillyard notes, it is often employed when one or more jurisdictions are uneasy about the perception of relinquishing control (2000, p. 101). Unified Command seeks to streamline a complex chain of command into a simpler, task-oriented structure.

Most commonly seen in the midst of wildland fire season in Australia and the western United States, the concept of Area Command allows for several simultaneous response efforts to effectively draw from a limited pool of qualified personnel and materials. The presumption is that if Area

Command must be implemented, multiple simultaneous incidents are taking place over a broad geographical area. Area Command is likely to involve individual ICs and IAPs, yet may most effectively be coordinated through a single, large-scale EOC. Truly multijurisdictional events drive its implementation, such as wildfires, flooding, and multiple terrorist acts.

### *Emergency Managers*

Now that a review has been made of the structure and layout of an emergency or disaster scene, emergency-management students or practitioners may be asking themselves, where do I fit in? The answer is, everywhere. A dearth of qualified and skilled emergency managers in most urban settings indicates that there will be more assignments than candidates.

Emergency managers very frequently end up within the Command and General Staff, often as ICs, LNOs, or PSCs. They will just as often be assigned to a frontline operations unit attacking a very specific threat or hazard. No two incidents unfold precisely the same way, even if there are commonalities. As a generalist, who likely also has areas of highly specialized training, a qualified emergency manager will be a very in-demand commodity during a crisis.

Finally, it is worth noting that the best way to select your assigned role is to be a part of the planning and preparedness process. The more involved an emergency manager is within his or her organization's crisis planning, the more likely that individual positions or functions will emerge which are particularly appealing or meaningful.

## AN URBAN PERSPECTIVE

The ICS, and more specifically NIMS, is an excellent tool for the urban emergency manager. But no system—no matter how well-designed or how broadly adopted—will magically solve all crises. NIMS represents a proven command framework, but it relies on personal and professional interactions to function. Competent and professional staffing of all roles, particularly those of the Command and General Staff, remains vital to success.

In densely populated areas, challenges are frequently magnified. And while NIMS unquestionably emerged from two disciplines that traditionally have a wide range of movement in the field—wildland firefighting and the military—urban emergencies are often constricted or vertical affairs. Even if needed resources are available, maneuvering them into proper position can sometimes be a difficult and lengthy process.

The core tenants of NIMS become all the more crucial in cities. As previously discussed, the density of civilians and critical infrastructure serves to magnify and accelerate the impacts of a given emergency of disaster. Losing electrical power and potable water to an acre in central Wyoming is a very different experience from losing power and water to an acre in central Manhattan. Quite simply, this requires urban emergency managers to be quicker and more flexible, particularly during the response phase.

These types of scenarios also place the focus back on planning and preparedness. It is absolutely vital that an effective emergency manager have both a strategic-level plan in the form of an EOP or similar document, and a tactical-level plan in the form of a current IAP. Certain core NIMS principles also gain importance in a high-tempo response. Management by objectives, a clear chain of command, and effective communications each become all the more critical in a dense urban setting.

Finally, multiagency drills and exercises are the best and most realistic training devices available to the emergency management community. Training in a classroom, participating in a tabletop exercise, and executing a full-scale field exercise are all crucial functions of an effective preparedness program. Such events are absolutely invaluable opportunities to introduce fellow planners and responders to the core concepts of incident command, and to one another.

## SUMMARY

The concept of ICS grew primarily from the experiences of the military and wildland firefighting communities. Effective command and control under extreme duress represented a true challenge, and key principles emerged as guideposts to a successful program.

In the United States, following the terror attacks of September 11, NIMS was adopted as the federal standard in emergency-management systems. Key principles to its successful implementation include effective planning and communication, clear chain of command, and logistics management. A range of corporate and nonprofit entities have also adopted the system. The accelerated tempo and heightened impacts of an urban environment mandates additional planning and preparedness efforts prior to an emergency.

## References

Erickson, P. (2006). *Emergency response planning for corporate and municipal managers*. Burlington, MA: Butterworth-Heinemann.

Haddow, G., Bullock, J., & Coppola, D. (2008). *Introduction to emergency management*. Burlington, MA: Butterworth-Heinemann.

Hillyard, M. (2000). *Public crisis management*. Lincoln, NE: Writers Club Press.

Pearson, C., Misra, S., Clair, J., & Mitroff, I. (1997). Managing the unthinkable. *Organizational Dynamics*, *26*, 51–64.

Purpura, P. (2007). *Terrorism and homeland security*. Burlington, MA: Butterworth-Heinemann.

## Further Reading

Department of Homeland Security (2008). *The national incident management system*. Washington, D.C: U.S. Federal Government.

Federal Emergency Management Agency. (2004). Retrieved from: <http://www.fema.gov/txt/nims/nims_ics_position_paper.txt>

Homeland Security Presidential Directive-5. (Revised, 2004). Washington, D.C.: U.S. Federal Government.

CHAPTER 7

# Collaboration in Urban Environments

## INTRODUCTION

As the sun rose over Louisiana and Mississippi on the morning of August 30, 2005, the world got its first real glimpse of the true destruction wrought by Hurricane Katrina. As had been widely reported, the city of New Orleans and the nearby parishes had at first thought to have come through the storm relatively unscathed as it came ashore the previous day.

Now, in the early morning light, the true scope of the devastation could be seen clearly. Several key levees in and around the city had suffered catastrophic failures. The storm surge—a wall of water pushed by the intense force of the hurricane-force winds—had traveled much further inland than was initially believed.

Hundreds of lives had been lost and ten-of-thousands more were in immediate danger. It became immediately, painfully clear that response and recovery was a job bigger than the federal government could handle alone. Collaboration with the private and nonprofit sectors was not a luxury, it was an absolute necessity.

As a concrete example, The US Army National Guard—the nation's home defense force—had a total of 7500 troops on active duty across the Gulf States when the storm came ashore. Within 2 weeks, the American Red Cross alone would deploy 75,000 trained volunteers to the region. Thus, a single volunteer organization activated 10 times the frontline personnel that the US military had in place on the eve of the disaster.

Federal resources, of course, poured into the region during this timeframe as well. Yet the point was made painfully clear: in a large-scale disaster, no government could not go it alone.



Urban Emergency Management. DOI: http://dx.doi.org/10.1016/B978-0-12-805307-2.00007-X

## ANALYSIS

### Some Context

As previously noted, the response to all disasters ultimately begins at the local level. Yet, it was not uncommon for local resources to be depleted. Across Europe and North America, this fact was arguably of the strongest motivators for the emergence of the federal system, where smaller entities such as cities, states, and provinces could seek aid and assistance via a coordinated regional or national system.

Sylves defines intergovernmental relations as the interaction of federal, state, and local officials and officials of the private and nonprofit sectors, as they collectively implement public policy (2008, p. 133). Yet, as the disjointed and uneven response to Katrina showed all too well, this is no simple task. Initially, ad hoc regional systems formed out of need and necessity, but these showed their shortcomings in large-scale incidents where trained personnel and materials were at a premium.

Ultimately, the concept of the incident command system (ICS) model has emerged as the glue to hold these disparate units together for a common objective. A broader recognition of the value if ICS began to emerge following wildfires and terror attacks during the 1990s, particularly among specialized operators. The wildland firefighting community continues to be the leading advocate of the system in the United States, similar to their counterparts in the Federation of the European Union Fire Officer Associations.

### Collaboration in Action

As is often the case with a new or emerging field or profession, emergency management has not been shy or timid about borrowing successful concepts from related functions. And there were multiple successful endeavors that limited the need to reinvent the collaboration wheel.

In an emergency-management context, collaboration may be defined as working with others in a joint intellectual or operational endeavor. Successful collaboration requires both horizontal and vertical integration.

#### *Collaborative Examples*

A number of existing operations exhibited traits and procedures from which emergency management could learn. Some common themes began to emerge, including best practices for ongoing joint efforts. Such collaboration requires communication and a common set of goals and objectives, which may be made more challenging in the midst of an emergency or crisis situation.

The modern hospital emergency room provides an intriguing example of collaboration. Prior to a patient even arriving at such an acute care center, the ambulance crew providing transport has likely forwarded their vital signs and overall condition to the destination. The patient will be met by a team of medical professionals who then perform a more thorough triage, and assign appropriate medicines, technologies, and subject-matter experts to the patient's care. For a person with a serious medical condition, this process can literally represent a life-or-death situation. The system has been refined over many years via operational experience and academic studies and performs at a high rate of success under very challenging conditions. Obstacles include maintaining appropriate staffing, and medicine and supply stockpiles—which is, as any medical professional will be willing to share, a constant fiscal and logistical challenge.

Military operations also require a highly refined level of collaboration. When deployed in the field in either exercises or combat operations, military units represent a highly interconnected and specialized system. As an example, take the infantry soldier, often referred to as the "tip of the spear" in military parlance. Before being deployed, he or she has been extensively trained as well as briefed on chain of command, opposing and friendly forces, and specific objectives. The modern infantryman carries not only state-of-the-art weaponry but also multiple communication devices and methodologies. He or she is likely backed by multiple air, sea, logistical, and cyber units (and facing the same). The ability to call on and interact with these supporting functions will have a great influence on the success or failure of attaining the stated mission objectives.

A final example of collaboration in action is provided by the wildland firefighting community. As noted in Chapter 6, Incident Command Systems and NIMS, the organizational systems employed to fight wildfires in the western United States have been hugely influential on the larger emergency-management community. The collaboration required to battle a large or fas-t-moving wildland fire is extreme. Such fires often start in isolated or hard to reach locations, making initial assessment and planning very difficult. A wildfire's ability to threaten people, animals, and structures often depends upon conditions outside of firefighters' control such as wind, temperature, and humidity. They must, therefore, be on offense and defense simultaneously—attacking the fire itself, while protecting their own personnel and equipment as well as civilians. Often spread out over a large geographical area, such incidents require adept and exacting coordination of assets ranging from individuals with pickaxes, to bulldozers and aircraft.

The emerging emergency-management and disaster-sciences communities closely examined each of these operations, and many more, to learn about interagency and intergovernmental collaboration. And the lessons learned have been incorporated into a strategic, high-level playbook known as the ICS.

### *The ICS/NIMS Model*

One of the key elements in the creation of the Department of Homeland Security (DHS) between 2001 and 2003 was that of effective coordination and collaboration. Federal defense, intelligence, law-enforcement, and emergency-management officials had received withering criticism for perceived failures prior to and during the September 11 attacks. A series of Homeland Security Presidential Directives, or HSPDs, set the tone for the new organization.

HSPD-5 clearly places the focus on vertical integration between governmental agencies rather than on the horizontal organization and alignment present in most other HSPDs. At the absolute core of HSPD-5 was the comprehensive incident-management system known as the National Incident Management System (NIMS).

As noted in Chapter 6, Incident Command Systems and NIMS, the system has become one of the pillars of emergency management within the United States. From its inception, it has sought to standardize disaster response across a wide range of threats and vulnerabilities as a true all-hazards model. Its simplicity and modular structure allow for an expandable or contractible command structure, allowing vertical and horizontal integration.

As described in the Introduction to the 2008 version, NIMS seeks "a systematic, proactive approach to guide departments and agencies at all levels of government, nongovernmental organizations, and the private sector to work seamlessly to prevent, protect against, respond to, recover from, and mitigate the effects of incidents, regardless of cause, size, location, or complexity." NIMS is thus not a truly tactical operational plan but rather a strategic and systemic guide to disaster response.

The core document addresses successful emergency management, breaking the ICS concept into 14 proven operational management characteristics:

- Common Terminology
- Modular Organization
- Management by Objectives
- Reliance on an Incident Action Plan
- Manageable Span of Control
- Predesignated Incident Mobilization Center Locations and Facilities
- Comprehensive Resource Management
- Integrated Communications
- Establishment and Transfer of Command

- Chain of Command and Unity of Command
- Unified Command
- Accountability of Resources and Personnel
- Deployment
- Information and Intelligence Management

The core NIMS document also outlines the command functions in the form of a single Incident Commander versus Unified Command, noting the essential differences between one emergency and a series of simultaneous emergencies.

## Defining Sectors

If collaboration and intergovernmental coordination includes multiple partners, it is vital to identify just who those partners might be. There are three major recognized sectors, and each is worthy of further review.

### *The Public Sector*

In general, the public sector may be thought of as government. But this may include a broader segment than it first appears. Consider that local, state, and federal public agencies are a part of this sector. In the United States, official tribal bodies would also be included. Virtually any entity that is funded with tax dollars and answerable to appointed or elected officials may be considered a part of the public sector.

At the local level alone, the sector may include cities, towns, villages, counties, and parishes. The state and federal levels likely represent multiple standing committees and occasional panels or commissions.

Perhaps more importantly, the public sector includes virtually all first responders. Local police, sheriffs, fire services, ambulance services, emergency managers, and public-works crews are all a part of the public sector. To simplify this broad subset, just remember that the first vehicle arriving on an emergency scene with a flashing light is virtually certain to be from a local public-sector agency.

If the incident is large or complex in nature, state or federal responders from law-enforcement and emergency-management agencies will also be deployed to the scene. These resources may also include military, National Guard, or homeland security units. Examples of public-sector entities would include:

- Local fire department hazardous-materials team
- County sheriff squad car
- State commission on flood-control levees
- National Guard debris-removal unit
- Federal law-enforcement agent/officer

### *The Private Sector*

By far the largest of the sectors, the private sector encompasses all for-profit and business entities. This would include not just corporations but also many professional trade groups, unions, associations, and personal networks. According to studies by DHS, the private sector owns or operates nearly 90% of critical infrastructure in the United States. If incidents occur within these facilities, employees of the private sector will be the de facto first responders.

Thus, size alone makes the private sector a crucial partner in the larger public-safety effort. However, an even more critical asset possessed by the private sector is subject-matter expertise. Envision the skilled operators of a 5-ton crane, an airfreight forwarder with intimate knowledge of aviation logistics, or the veteran mechanical engineer with an expertise in high-volume water pumps.

Finally, the private sector possesses an enormous amount of capital in the form of cash, goods, and liquid assets. Large-scale disasters require significant amounts of funding for an effectively coordinated response and recovery effort. In a capitalist world, products and services are expedited most effectively by being effectively and reliably funded.

Yet, the private sector should not be considered a collection of only the wealthiest companies and individuals. In the 21st century, private citizens can band together to financially support urgent needs via online and Internet platforms such as GoFundMe and Kickstarter—and this is a trend likely to continue. Examples of diverse private-sector entities would include:

- Physical security manager for Bayer AG
- Crane operator for XYZ Construction Company
- Chief Financial Officer for JPMorgan Chase
- Maintenance Foreman for Caterpillar Incorporated
- Cargo Loadmaster for United Parcel Service

### *The Nonprofit Sector*

If the public sector is the first to come to mind, and the private sector is the largest, then the nonprofit sector is the most underappreciated. The nonprofit sector represents charities and not-for-profit corporations that offer crucial expertise, materials, and support during crises. These largely volunteer groups play an ever more-critical role in emergencies. The National Volunteer Organizations Active in Disaster (NVOAD) is a collective of dozens of groups—58 as of 2016—possessing various response and recovery capabilities. Many of these individual organizations are faith-based, and see such volunteer work as an important piece of their community-outreach duties.

A key part of their services is training, ranging from basic CPR/AED classes to complex and high-level mass-care and sheltering-management coursework.

The nonprofit sector may include loose confederations of short-wave radio operators as well as formal, international charitable organizations.

During disasters, these organizations offer an incredibly broad set of support functions, including emergency rations, clothing, housing vouchers, canteen services for responders, counseling services, and even laundry. Such groups are absolutely critical partners to emergency managers in cash-strapped urban agencies. Examples of nonprofit entities include:

- Salvation Army canteen truck driver
- Red Cross shelter facility
- Médecins Sans Frontières (Doctors Without Borders) medical tent
- United Methodist Volunteers in Mission clean-up crew

### *Theory Meets Practice*

Translating planning and organization from a static document into actual practice is always a challenge. Erickson refers to the current cross-sector efforts at emergency management nothing less than a paradigm shift, noting that the new model "cuts across long-established boundaries of jurisdictional authority, forcing not only a significant increase in interagency cooperation but also a growing partnership among federal and state authorities as well as private-sector organizations and corporations" (2006, p. 232).

Erickson also notes a model emerging across much of Asia—one that holds corporations largely accountable for their own initial response and recovery efforts, and costs. He cites multiple governments following the dictum of "those who are responsible for causing hazards and risks are precisely the persons who have primary responsibility for controlling those hazards and risks" (2006, p. 232).

The Deepwater Horizon oil spill in April of 2010 does make for a slightly murkier picture in the United States. The federal government focused primarily on public notification and clean-up efforts during the disaster, and expected the private-sector owners and operators of the crippled oil-drilling platform to stop the actual leak. Multiple efforts to seal the underwater oil well responsible for the largest spill ever in US waters took nearly 90 days to complete. Ultimately, the lessee of the oil platform, BP, suffered far more dire financial consequences than the owner of the rig (Transocean) or the primary drilling contractor (Halliburton).

The Deepwater Horizon incident ultimately represents a lack of planning, effective governmental oversight, and preincident collaboration. The resulting response was inefficient and poorly coordinated. Once again, the previously referenced lack of imagination to predict the raw scope and scale of the emergency was a critical factor. To some extent, all of the stakeholders either never considered that a major leak could occur on the seabed, or presumed that one of the other organizations had the resources and knowledge to respond to and mitigate such a threat. Whether included in the planning or not, the initial cooperation across sectors will largely depend on interpersonal relationship among stakeholders.

By contrast, the successful collaborations within hospitals, the military, and wildland fire operations exist—at least in part—due to well-defined common objectives and a level of trust among participants. What cannot be easily or accurately captured on paper is the core ability to complete a primary function, yet be flexible and reliable enough to go just a bit above and beyond on behalf of fellow stakeholders.

## AN URBAN PERSPECTIVE

In densely populated areas, emergency-management challenges are magnified. Urban emergencies are often constricted or vertical affairs. As noted in Chapter 5, Assessing Risk in Urban Areas, even if needed resources are available, maneuvering them into position can sometimes be a difficult and lengthy process to accomplish.

The core tenants of incident management become all the more crucial in cities. As previously discussed, the density of civilians and infrastructure serves to magnify and accelerate the impacts of a given emergency of disaster. Ultimately, this requires urban emergency managers to be more decisive and more flexible, particularly in the response phase. In addition to implementing an established incident-management system such as NIMS, there are several collaborative tools available to emergency managers before, during, and after a crisis.

### Tools of the Trade

Later chapters focused on the individual phases of emergency management will address the specifics of prevention, preparedness, mitigation, response, and recovery. However, a few key collaborative elements straddle multiple phases of emergency management. Combining resources from the public, private, and nonprofit sectors often results in a force multiplier effect in which the resulting impact is greater than the sum of its parts.

#### *Emergency Operations Center*

The emergency operations center, or EOC, is a core component of the command and control function during an emergency. Slightly removed from the core of the incident, the EOC provides strategic guidance and high-level support to the entire Command and General Staff. An EOC manager may be designated to run the actual facility, yet a senior staff member will always be present—often the Incident Commander or his/her designee. Multiple Section Chiefs will also run their operations out of the EOC.

In recent years, there has been some degree of professional controversy regarding whether members of the private and nonprofit sectors should be collocated in an EOC. The answer is largely dependent upon the emergency itself.

A short-term incident in which public-sector responders and public-works employees can address the situation would likely not require the presence of other sectors in the EOC. An example may be a hostage-barricade situation in a public building. As a law-enforcement incident, an EOC may be stood up to support public-sector response and recovery efforts, but impacts to the private and nonprofit sectors are likely to be minimal.

Conversely, if an incident impacts the private and nonprofit sectors, or is expected to last for an extended period of time, the presence of those sectors in the EOC should be welcomed and encouraged. For example, imagine a major water main collapse in the midst of an urban center. Flooding, water damage, and traffic interruptions will affect multiple sectors, and the response and recovery timeframe may be extensive. The presence of nonprofit and private-sector representatives in an EOC may prove highly beneficial.

A key to successful implementation is that members of the private and nonprofit sectors must receive adequate orientation and training to the EOC prior to an emergency. If "outsiders" are brought into the EOC for the first time in the midst of a crisis, they are being set up to fail. If, however, they have an understanding of the EOC and core incident-management concepts, these representatives can be vital force multipliers by acting as liaisons to and from their respective sectors. Private- and nonprofit-sector representatives should be familiar with both the layout and the technology utilized in the EOC. From an administrative standpoint, sector representatives should meet the same standards as their emergency-management counterparts in terms of expertise, temperament, and background check requirements.

### *Joint Information Center*

A joint information center, or JIC, can be envisioned as an EOC for media relations. If there are multiple agencies or sectors responding to a crisis, unified messaging becomes critical. It is vital that all sectors, and the public, be receiving clear and unambiguous information.

The most crucial tool for this process is the JIC. As the center for crisis communications during an incident, the JIC is a physical or virtual location where the public information officers and crisis communications officers of stakeholders meet and discuss messaging.

A primary function of the JIC will be to dispel rumors and to provide updates from the incident scene and/or the EOC. The JIC may also be utilized as a gathering point for media representatives and reporters and may be the site of press conferences. Any physical JIC location should be somewhat geographically removed to keep the media from interfering with either the response location or with the EOC.

An effective JIC may also manage alert mechanisms geared to the public, including mass messaging, social media, and formal press releases. When executed correctly, the JIC concept can be an absolutely vital partner across all impacted sectors.

### *Drills and Exercises*

The maxim "practice makes perfect" has been around for many, many years—and with good reason. Perfection may not be an attainable goal in the chaotic world of emergency management but improvement certainly is.

Any truly collaborative cross-sector effort will include a strong training and exercise program. As noted in the prior assessment of the EOC concept, it is vital that participants know their roles and functions prior to a true emergency incident occurring. An established training and exercise regimen has a proven track record of success in this arena (Fig. 7.1).

The most effective methodology for emergency managers to coordinate cross-sector drills and exercises is that of progressive complexity. This concept relies upon a series of gradually more complicated drills and exercises for

**FIGURE 7.1**
A high-rise evacuation full-scale exercise in Thailand, utilizing both public-sector and private-sector participants. *Credit: Shutterstock.*

participants. It may begin with a tabletop exercise in which participants are guided through a series of scenarios and injects—adjustments and occurrences within the larger incident—sometimes referred to as a discussion-based exercise. Next, participants are involved in increasingly complex operations-based exercises, perhaps two or three based upon the same hazard or threat. Ultimately, actual units would be deployed to the field in a simulated response, while all supporting mechanisms (EOC, JIC, etc.) would also be activated.

The DHS Homeland Security Exercise and Evaluation Program (HSEEP) is an excellent example of formalized and vetted exercise design. While its stringent structure does require an extensive degree of forethought and planning that not every jurisdiction or organization can attain, it is the standard against which all holistic exercise programs are measured.

### *Formal and Informal Organizations*

A number of professional organizations exist that assist emergency-management professionals in collaborating with one another. Just two will be examined here, but the author strongly encourages students and practitioners to conduct their own surveys to determine which such groups may be active in their area.

The International Association of Emergency Management (IAEM) is the largest formal network of professional emergency managers, with approximately 7500 members across the globe. The association's mission statement is "to serve its members by providing information, networking, and professional opportunities and to advance the emergency-management profession."

The group also hosts a large annual conference with multiple educational tracks and oversees the rigorous, peer-reviewed certification known as the Certified Emergency Manager (CEM) designation. (Full disclosure, the author does hold the CEM designation.) The organization also produces two electronic newsletters, the weekly *IAEM Dispatch* and the monthly *IAEM Bulletin*. Regional conferences are also often held in Asia, Europe, and the Middle East.

The Big City Emergency Managers, Inc., is an association of—not surprisingly—emergency managers from big cities. Perhaps less formal than the IAEM, the group coalesced in 2005 as leaders from multiple American jurisdictions recognized that large cities faced many of the same fiscal and operational challenges. The need was real, and the group quickly evolved from 4 to 15 member jurisdictions, and now represents fully 30% of the US population.

The group members share an active network of phone and e-mail exchanges and meet approximately twice a year to discuss emerging issues and solutions. The group's mission statement is "to foster the development and growth of robust and nimble emergency-management operations in the nation's largest, most at-risk metropolitan jurisdictions so that the country is better positioned to prevent, protect against, prepare for, respond to, and recover from major incidents and catastrophic emergencies." Member jurisdictions include:

- City of Boston
- City of Chicago
- City of Dallas
- City and County of Denver
- District of Columbia
- Harris County (Houston)
- City of Houston
- City of Los Angeles
- Miami-Dade County
- Newark-Jersey City Urban Area
- City of New York
- City of Philadelphia
- County of San Diego
- City and County of San Francisco
- City of Seattle

As referenced in Chapter 1, Overview of Urban Emergency Management, and Chapter 2, Development and History of Emergency Management, such professional networks indicate the emergence and growth of the profession as a whole. And yet true collaboration requires casting a much wider net.

### *Joint Committees*

They may have multiple titles—committees, commissions, task forces—but cross-sector working groups can be a vital part of an urban area's planning and mitigation capabilities. Each jurisdiction or organization will experience variations on the theme of such groups, yet when done correctly, they can have immense impact.

As a best practice, any large urban area should have a standing group responsible for reaching across sectors to gather the subject-matter expertise found within. For such a committee or task force to be truly successful, it will need leaders in multiple sectors to champion its efforts.

The most successful joint committees or task forces effectively act as a network of networks or an association of associations. This trait goes back to the force multiplier concept, in which each participant is, in turn, acting as a conduit to a number of contacts and associates within his or her profession. For example, if the head of security for a large hotel is participating in the group, he or she is not representing a single hotel but rather *every* hotel and resort within the urban area. In such an empowered liaison function, the role then acts as a conduit to and from a much larger constituency. To be most effective, individuals must possess the mentality of a true team player.

Such organizations may be difficult to first form, yet seem to succeed when the right personalities from a broad range of concerns are at the table. Participation should be seen as a privilege and an honor rather than as a right or a burden. Finally, cross-sector committees or task forces must be task-oriented. Specific projects or goals are vital to the continued success and relevance of such groups, which must not be allowed to devolve into a "meeting for the sake of meeting" status.

The case study at the conclusion of this chapter focuses on one such organization seeking to empower its members in just such a manner.

## Emergency Manager Role

Effective collaboration can be a literal lifesaver in times of crisis. This chapter has examined a number of potential partners, tools, and resources. Ultimately, it will be the role of an effective urban emergency manager to bring these elements together.

Emergency managers may find a NIMS-specific role within the Command and General Staff, or be assigned to a frontline Operations unit attacking a very specific threat or hazard. As both a generalist and a specialist, a highly qualified emergency manager will always be in demand during a crisis.

But the most important role that an emergency manager can play is that of Chief Collaborator—the connective tissue that holds together disparate interests and stakeholders before, during, and after a disaster. It is almost never easy, and rarely publicly acknowledged. And yet this liaison role fills a critical void between other professionals, and in the best of circumstances, will motivate them to align behind a common set of goals and objectives.

# CASE STUDY: THE CHICAGO PUBLIC-PRIVATE TASK FORCE

## Background

Early in 2010, the city of Chicago recognized the need to form a collaborative partnership including both private-sector and public-sector organizations. Initially called the Critical Infrastructure Resiliency Task Force (CIRTF), the group concentrated on issues of preparedness and business continuity. Regular meetings focused on analyzing and discussing systemic threats which could impact multiple sectors.

In 2012, the jurisdiction hosted the North Atlantic Treaty Organization (NATO) summit, which included senior elected and defense officials from approximately 25 nations. In addition to logistical and terrorism concerns, the summit attracted a wide range of protesters, providing some degree of disruption to multiple entities in the city's central business district.

Extensive planning and preparation had taken place prior to the summit, including multiple threat-specific, cross-sector committees. It became clear that the standing CIRTF working group focused on critical infrastructure issues within the preparedness and response phases needed to be broadened. Given the overall success of the planning effort, and after much consideration by senior participants, the scope and scale of the working group was expanded, and its role and title were redefined.

With that revision, the Chicago Public-Private Task Force (CPPTF) was born in late 2012. Its initial mission statement was to "provide an ongoing and consistent forum for information sharing, planning, and building resilience between and among Chicago public-safety agencies and operational private-sector partnerships." From the beginning, the group was envisioned as a network of networks—an assembly of individuals who could represent much larger numbers within the public, private, and nonprofit sectors.

Somewhat uniquely, the task force was established under a cochairmanship—one chair would rest with the public sector, and the other with the private/nonprofit sector. As the group was originally organized by the city's Office of Emergency Management and Communications (OEMC), the executive director of the department held the inaugural public-sector chair. The other chair was held by the ChicagoFIRST organization, a group consisting primarily of resiliency, business-continuity, and security experts from within the financial sector.

Other nonprofit and private subsectors which were represented included building owners and managers, colleges and universities, cultural properties, hospitals and medical providers, hotels and tourist destinations, and retail and store owners. On the public-sector side, representatives included traditional responders such as fire, police, and emergency management as well as DHS, public health, and the city's buildings department.

## Best Practices

Expanding on the earlier CIRTF template, the task force established a number of organizations as permanent members of the group and is empowered to bring in specific subject-matter experts from other sectors and subsectors as needed. Meetings of the full task force are scheduled with monthly frequency, with subcommittees or working groups meeting on an as-needed basis. Attendance and activity criteria were also set for continued membership, assuring that no group would benefit from membership without making contributions to the larger effort as well.

With the structure in place, the CPPTF got to work on a number of potentially thorny, cross-sector issues. Specific projects include:

- A credentialing program for critical infrastructure owners and operators to reenter their facilities following a disaster
- An EOC training program for representatives of the private and nonprofit sectors
- Reviewing and implementing incident-command and incident-reporting software programs and systems
- Education programs and presentations for residents and commuters on complex topics such as evacuation and shelter-in-place planning
- Cross-sector common operating picture (COP) best practices

To increase participation and inclusiveness, the task force also rotates its monthly meeting location between public-sector and private-/nonprofit-sector locations, and members provide brief updates and overviews of their organizations to enlighten fellow participants about their capabilities and areas of expertise. The result has been not only a formal task force but also an informal network of networks that frequently contact one another independently and directly regarding topics of mutual concern.

As a specific example, the CPPTF has also been instrumental in exercising and drilling concepts such as just-in-time credentialing for stakeholders to reenter facilities following a disaster. Several members met in the field to test a simple, mobile solution to a very challenging aspect of business-continuity planning. The task force had determined that utilizing a preexisting list of key operational managers for key infrastructure locations would be preferable to a more expensive and complex "hard card" credentialing program. The vetted list of eligible stakeholders was managed by the local law-enforcement fusion center and could be deployed to the incident perimeter regardless of the status of electrical power or Internet connectivity.

As of 2017, the task force is still highly active, and is considering a new and expanded list of projects, subcommittees, and potential permanent member organizations. The group has also transitioned new leaders into both the private-sector and public-sector cochair seats as it continued to grow—an excellent sign of likely longevity. While no such group can solve every problem faced by a large and complex urban area, the CPPTF stands as a successful model for ongoing collaboration among public, private, and nonprofit organizations in a modern city.

## SUMMARY

Working across public-, private-, and nonprofit-sector boundaries is becoming a requirement in an era of reduced budgets and increased expectations. Modern hospital emergency rooms, specialized military units, and wildland firefighters all provide valid examples of successful collaborative efforts. Effective contributions from all sectors will be vital during and after a catastrophic event, as was all too clearly displayed by the examples of Hurricane Katrina in 2005 and Deepwater Horizon in 2010.

Cooperative tools available to emergency managers include the EOC, the JIC, joint training programs, and planning committees and task forces. A cross-sector task force in Chicago is cited as an example of effective efforts at collaboration in an urban setting.

## References

Erickson, P. (2006). *Emergency response planning for corporate and municipal managers*. Burlington, MA: Butterworth-Heinemann.

Sylves, R. (2008). *Disaster policy and politics*. Washington, D.C: CQ Press.

## Further Reading

Big City Emergency Managers, Inc. (2016). Retrieved from: <www.bigcityem.org>.

Department of Homeland Security (2008). *The national incident management system*. Washington, D.C: U.S. Federal Government.

Department of Homeland Security. (2013). HSEEP Program. Retrieved from: <http://www.fema.gov/media-library-data/20130726-1914-25045-8890/hseep_apr13_.pdf>.

Federal Emergency Management Agency. (2004). Retrieved from: <http://www.fema.gov/txt/nims/nims_ics_position_paper.txt>.

Haddow, G., Bullock, J., & Coppola, D. (2008). *Introduction to emergency management*. Burlington, MA: Butterworth-Heinemann.

Homeland Security Presidential Directive-5. (Revised, 2004). Washington, D.C.: U.S. Federal Government.

International Association of Emergency Managers. (2016). Retrieved from: <www.iaem.com>.

CHAPTER 8

# Emergency Management and Homeland Security in Urban Areas

## INTRODUCTION

### Brief Case Study

On April 15, 2013, two explosions rang out near the finish line of the renowned Boston Marathon. Acts of terrorism within an urban area, the blasts claimed the lives of three civilians—Krystle Campbell, 29, Lu Lingzi, 23, and Martin Richard, 8—and injured more than 250 others, some grievously. The explosions set in motion a massive response involving law enforcement, emergency medical services, and emergency management. Images of the perpetrators, two brothers who were naturalized American citizens of Chechen/Kyrgyzstani heritage, were revealed during an intense investigation that leveraged witness statements and vital video surveillance footage from both the public and private sectors.

Shortly after the photos were released to the media and to the public on April 18, the perpetrators ambushed and killed Massachusetts Institute of Technology police officer Sean Collier, starting a chaotic 24-hour period that involved a furious manhunt—including a shootout with police on a suburban residential street in which the perpetrators used pipe bombs as grenades and an sport utility vehicle as a battering ram—and ultimately the capture of one suspect and the death of the other. The interim represented an unprecedented lockdown of a major American city, with law-enforcement and emergency-management officials recommending that civilians shelter in place and eliminate all nonemergency travel. Public transportation systems and most public institutions were closed, creating an eerie lack of pedestrian and vehicular traffic in a major urban center.

Investigation revealed that the initial blasts had been caused by so-called pressure cooker bombs placed in backpacks, created by packing the cooking vessels with black gunpowder and shrapnel. The devices were relatively easy to construct, with plans readily available on the Internet and advocated by foreign terrorist organizations such as Al Qaeda in the Arabian Peninsula



Urban Emergency Management. DOI: http://dx.doi.org/10.1016/B978-0-12-805307-2.00008-1

(AQAP). While the older sibling had traveled to areas of Chechnya with known extremist Islamist elements, the brothers appeared to have been radicalized largely via the Internet. Their violent actions led to an unprecedented public-safety response in a metropolitan area. It was to be, unfortunately, a painfully clear harbinger of things to come for urban emergency managers.

### Context

Terrorism has been with us for centuries, at least as long as the reality of standing armies. From the time that a group felt overwhelmed or oppressed by a superior military force, the concept of asymmetrical warfare—including the tactic and methodology of terrorism—has flourished in many parts of the world.

In the Middle East in particular, thousands of years of overlapping and competing populations have resulted in groups resorting to such tactics, from the Sicariis pitted against the Romans to Shia assassins targeting the Sunnis, and from the Irgun attacking the British to Hamas targeting Israelis.

And yet, such tactics are in no way limited to such a small swath of the planet. In the past decade, asymmetrical attacks have resulted in fatalities on every populated continent. In terms of holistic, all-hazards planning, such threats are a part of the larger human-caused hazard spectrum, particularly in large cities.

The private sector is also keenly aware of the necessity to mitigate losses from physical violence, as a 2004 *Business Insurance* piece titled "Natural, Manmade Disasters Require Similar Risk Planning" makes clear. The article's author observes that "mitigation and business-continuity strategies . . . developed for natural disasters share some similarities with preparations for potential terrorism events," and notes that Liberty Mutual has found "that its clients increasingly want terrorism and crisis management evaluations as part of" their companywide risk assessments (2004, p. 31–32). A modern emergency manager in an urban area must also anticipate terrorist incidents impacting his or her organization or jurisdiction.

## ANALYSIS

### Terrorism and Homeland Security

To understand such a complex issue, it is often best to begin by defining it. The most commonly cited definition of terrorism is that provided by the US federal code and utilized by the Department of Justice (DOJ): "The unlawful use of force or violence against persons or property to intimidate or coerce a government, the civilian population, or any segment thereof, in furtherance of political or social objectives" (C.F.R., 1969).

Terrorism is therefore the use of force, or threatened use of force, to achieve a societal objective. There are a myriad of groups that fit this definition, and federal law enforcement breaks them into two large categories, domestic and international. Most urban emergency managers are familiar with both types, their jurisdictions having been not only a target but also likely hosting individuals plotting to carry out attacks elsewhere.

Indeed, the US federal government issued its first presidential disaster declaration due to terrorism following the 1993 truck bomb international terrorism attack on the World Trade Center in New York City. This was followed by a directive from then-president Bill Clinton in 1995, following the Oklahoma City domestic terrorism truck bombing, which declared the DOJ to be in charge of crisis management and the Federal Emergency Management Agency (FEMA) to be in charge of consequence management. In fact, the terms crisis and consequence are still widely utilized by the DOJ today, and roughly correspond to the response and recovery phases, respectively.

And while the use of such terrorist tactics ebbs and flows over time, the beginning of the new century saw spasms of violence in new locations. Soon after the September 11 attacks in the eastern United States, coordinated attacks took place in a number of other metropolitan areas, including Madrid (2004), London, (2005), and Mumbai (2008). Large population centers in democratic nations were now fair game in terms of terrorist targeting. Not since the "bad old days" of the Irish Republican Army and Black September had cities seen such ideological violence. Governments scrambled to implement (or reimplement) policies and procedures to tighten security within urban centers (Fig. 8.1).

**FIGURE 8.1**
The blending of emergency-management and homeland-security functions has led to additional cross training and joint deployments such as these FDNY and NYPD vessels near Manhattan. *Credit: Shutterstock.*

In addition to structural and administrative changes, extensive updates were made to the planning process. As noted by Bevelacqua and Stilp, "Emergency responders including fire departments, law enforcement, emergency medical systems, hospitals and emergency management must have plans in place and must be ready to respond when these incidents occur" (2009, p. 1–3). It is now standard procedure for large American cities to have an advance threat assessment coordinated by local law enforcement for any large events and to build operations plans based upon such assessments. Such efforts involve the entire public-safety community, rather than just law enforcement. The previously referenced "See Something, Say Something" campaign provides a key example.

Terrorism also has made the emergency responder community in many Western cities "flatter" and less hierarchical. Individuals coordinate frequently across department and agency lines—for example, it is commonplace to see a police Bomb and Arson Squad training alongside a fire services Hazardous Materials Team. Emergency managers, firefighters, and paramedics are much more likely to be receiving and reviewing updates from fusion centers and similar intelligence distribution systems.

Homeland security can thus be viewed as the whole of this system aligned to counter the threats posed by domestic and international terrorism and to mitigate the resulting impacts of any attacks that do take place. The US government defines homeland security as a concerted national effort to prevent terrorist attacks within the United States, reduce America's vulnerability to terrorism, and minimize the damage and recover from attacks that do occur (DHS, 2002). This definition is particularly telling in that it spans all phases of emergency management—prevention, preparedness, mitigation, response, and recovery.

## Key Similarities and Differences

The Sylves definition cited in Chapter 1, Overview of Urban Emergency Management, is again relevant. Emergency management is the discipline and profession of science, technology, planning, and management to address extreme events with the potential to injure or kill large numbers of people, do extensive property damage, and disrupt community life (2008, p. 5). There is, therefore, a significant overlap between the fields of homeland security and emergency management.

Each is ultimately a public-safety function, protecting civilians and critical infrastructure from harm. In the United States, both fields also utilize the National Incident Management System (NIMS) command structure. Both areas of expertise are also largely populated by former military, law enforcement, and fire services personnel. As relatively new fields of study, emergency management and homeland security have by any measure experienced rapid growth over a short period. As a tangible example, no homeland security degrees were offered by accredited colleges and universities at the turn of the century. An article in *The Atlantic* identified 308 such degree programs as of 2016 (Brill, p. 64).

In the United States in particular, this intertwined relationship has created some tension on both an academic and operational level. Academically, homeland security would appear to be an element of the larger, broader emergency-management function. Operationally and organizationally, however, FEMA is now a part of the larger, 240,000-employee strong Department of Homeland Security (DHS). While there has been some argument to make FEMA a permanent cabinet-level agency with greater autonomy—particularly in the wake of the botched Hurricane Katrina response in 2005—no such change has occurred.

The argument to return the FEMA czar to prominence rests upon a key difference between emergency management and homeland security. The former focuses upon the all-hazards model of risk management, while the latter focuses exclusively on intentional, human-caused acts—the nexus of terrorism. The FEMA management model could be said to be more inclusive or coordination oriented, while the DHS management model is more military or command-and-control in nature.

It is key to note that operators and practitioners often see great overlap between different causations. For example, a mass-casualty incident as a result of an accidental explosion at a chemical plant will require much the same mitigation and response effort as an intentional release of similar chemicals by terrorists. Thus, while tension between the two emerging fields certainly continues, there is an acknowledgment at both the academic and operational levels that they certainly share more commonalities than differences.

A dynamic shift in federal funding also took place following the September 11 attacks. Historically, FEMA distributed approximately $175 million annually to its state and local partners (Haddow, Bullock, and Coppola, 2008, p. 305). By contrast, between fiscal 2002 and 2015, DHS distributed approximately $40 billion dollars via state and local grants (Brill, 2016, p. 67), representing more than a 15-fold increase over the prior funding levels. Inarguably, due in large part to public perception, the federal funding spigot opened wide following the 2001 Al Qaeda attacks.

Large municipalities in Europe face an even greater challenge. As the spate of terrorist attacks across France and Belgium in 2015 and 2016 graphically displayed, there is no shortage of disaffected and willing young men (the attackers were all male) currently living on the continent. This status is due to many complex factors, including the relative lack of assimilation of minority populations into the larger European society and to the global economic recession at the turn of the decade, which hit Europe especially hard. When relative isolation and high employment are mixed with the allure of slick Islamic-extremist online propaganda, the results can destroy many lives.

Emergency and crisis managers across Europe—including those in corporate, nonprofit, and governmental organizations—were very quickly thrown into a similar situation to that which their North American counterparts had experienced just a few years earlier. The system had underestimated the overall threat posed by terrorism and was rapidly transforming to catch up. Federal spending across the continent spiked, and law-enforcement units became more militarized in appearance and function. Active-duty military troops were deployed to transportation hubs and city centers, and emergency managers were expected to be quick studies in learning the basics of antiterrorism and counterterrorism policies and procedures.

## Antiterrorism, Counterterrorism, and Beyond

Two terms frequently cited in any conversation comparing and contrasting emergency management and homeland security are antiterrorism and counterterrorism. Like the larger fields themselves, the two functions overlap and align, yet offer distinct differences.

Antiterrorism may be thought of as efforts taken to discourage, deter, and mitigate potential acts of terror. Threat-specific analysis, deterrence messaging, target hardening, and mitigation planning may all be viewed as functions of antiterrorism.

Counterterrorism is different in that it involves active or kinetic actions to prevent or respond to acts of terror. Overseas drone strikes and special-forces operations, as well as domestic tactical teams, may be viewed as direct counterterrorism functions.

In the days of civil defense being a primary safety function in Western democracies across Europe and North America, the presumption was that a known opponent—the Soviet Union and its Warsaw Pact allies—would be the aggressor against the Unites States and its North Atlantic Treaty Organization (NATO) allies. This necessitated a controlled domestic response to what was anticipated to be catastrophic damage should the so-called Cold War devolve into a Hot War. It was also presumed that there would be a geopolitical crisis that would provide substantial warning prior to any such instance. But times have changed.

The Islamic State in Syria and the Levant (ISIL)—also known as ISIS, Daesh, and the Islamic State—represents a severe and radical escalation over their predecessors. While Al Qaeda was violent and hell-bent on inflicting large-scale attacks on Western targets, the new breed of Islamic extremism takes that violence to a pathological level not previously seen. Their deliberate targeting of civilians, and horrific executions of prisoners within the swath of territory they have occupied, is matched only by their relentless online marketing and recruiting efforts.

Today, in part due to such extremism, emergency managers in major urban areas are likely to participate in antiterrorism efforts much as they would take part in fire-education programs or severe-weather mitigation operations. Such involvement is commonplace as terrorism is often considered one of many threat vectors in terms of all-hazards emergency management. In this new reality, intentional, no-notice, human-caused incidents must be taken into consideration as a part of any detailed risk or threat assessment.

## AN URBAN PERSPECTIVE

Given terrorism's stated goal of creating fear and chaos within a target population or subgroup, it is no surprise that the majority of modern terror attacks take place within highly urbanized areas. Assessing the current threat picture and implementing proven emergency-management practices are valuable tools.

### The Current Threat

In many ways, terrorism represents a unique hazard to the urban emergency-management community. Terror, by definition, indicates the presence of a cognizant antagonist with the ability to resist and counter attempts to effectively prevent or mitigate an attack. In contrast to a structural collapse or a severe storm, a terrorist really is trying to kill you and your fellow responders, and thus represents an adaptive threat.

As noted in Chapter 4, Human-Caused Hazards and Hybrid Hazards, the tactics and processes utilized by terrorists continue to evolve and metastasize. They include:

- Assassinations or targeted killings
- Bombs, incendiary, or explosive devices
- Chemical, biological, or radiological releases
- Cyber attack

- Hijacking
- Kidnaping or hostage-taking
- Sabotage
- Small-unit or group assault
- Vehicle as a weapon

The world in which we live necessitates that any response to an incident must presume terrorism until proven otherwise. First responders and supporting units are often targeted by terrorists, and emergency managers should not deploy without considering secondary devices or secondary attacks as a very real threat. Such abundant caution should also be included in any and all multidepartment or multijurisdiction response plans.

### *International Terrorism*

A major category of terror activity is international in nature. International terror may be defined as acts of terrorism that transcend established national borders. Planning may take place by individuals in one nation, with terrorists training in a second nation, funding coming from a third nation, and the actual attack intended to take place in a fourth. This description applies precisely to the aviation-based attacks by Al Qaeda in 2001 and to the Islamic State in Iraq and the Levant (ISIL) attacks on Brussels and Paris in 2015 and 2016.

State-sponsored terrorism is an important subcategory of international terrorism. Many national governments see terror as simply another weapon in their military or diplomatic stockpile; a method to achieve their nationalistic goals. While more common in the late 20th century, an ongoing example of state-sponsored terrorism would include Iran's support for the radical Shiite organization Hezbollah. The US Department of State's annual *Country Reports on Terrorism* is an invaluable resource for professional emergency managers seeking to better understand such associations.

### *Domestic Terrorism*

Another major category of terror activity is domestic. Domestic terror is defined as acts of terrorism that occur within a single, established national border. Planning, funding, and the attack itself all take place in a single country. Such groups often include single-issue extremists with far-left or far-right ideologies. In the United States, groups such as the Earth Liberation Front and the Sovereign Citizens movement would be considered domestic terror in nature. In 2014, after an inexplicable multiple-year lull, the Federal Bureau of Investigation (FBI) reconstituted its domestic terror unit to counter such threats.

The 1995 Oklahoma City truck bombing is an example of domestic terrorism in the United States. Perpetrated by a veteran of the US military who had adopted extremist antigovernment views with a small circle of supporters or sympathizers, the attack killed 168 people, including 19 children. The incident remains the largest domestic terrorism attack in US history.

### *Lone Wolves*

Antiterrorism and counterterrorism officials continue to be concerned with the emerging phenomenon of so-called lone wolves. A lone wolf terror suspect could be loosely described as an individual with minimal formal training or outside contact with others. While they may harbor radical and violent ideologies, this relative lack of contact with outsiders makes them more difficult for intelligence services and law enforcement to detect during the preattack planning phase. Such attacks are generally smaller in scale with fewer fatalities than a major attack with multiple participants, yet they represent a true threat to public safety.

The mass shooting that took place in June of 2015 at the Emanuel African Methodist Episcopal Church in Charleston, South Carolina, is an example of a so-called lone wolf. Nine civilians were killed in a premeditated attack by an adherent of the virulently racist Christian Identity ideology. The attacker stated that he sought to start a race war with his actions, survived the mass shooting, was charged with 33 federal hate crimes and sentenced to death.

### *Hybrid Motivations*

And while it is tempting to put each violent radical into a nice, neat motivational "bucket," this is simply not realistic. Just as hybrid disasters now join the more traditional natural and human-caused categories, hybrid motives have also emerged in acts of terrorism.

There is strong emerging evidence that individuals suffering from mental illness or a personality crisis are much more susceptible to the types of online and Internet recruiting and propaganda perpetuated by ISIL and similar radical organizations. While these individuals may very well have lashed out regardless, the slickly packaged and alluring messages of violence and extremism make such actions all the more likely. Researchers, prosecutors, and law enforcement will continue to struggle to find identifiers prior to such radicalization leading to violent actions.

An example of this hybrid type of terrorist action would be the 2016 attack on the Pulse nightclub in Orlando, Florida. The attacker had espoused sympathies for radical Islamist groups in the past but had also suffered from failed relationships, had a history of domestic violence, and may have been struggling with defining his own sexuality. All of these factors certainly contributed in some degree to his violent shooting rampage that left 49 civilians dead, but because he perished in the attack as well, we may never know the exact influence or motivation provided by each.

## Best Practices

Ultimately, the effective management of emergency-management and homeland-security functions must be addressed at a federal or national level. While the initial planning and response will begin at the local level—including all of the previously citied terror incidents—overarching goals and objectives are a necessity. Multiple documents have been drafted and released in recent years to address the planning and coordination shortfalls made evident by the September 11 attacks.

In the United States, the DHS National Prevention Framework, released in 2013, notes seven core capabilities for avoiding or minimizing acts of terror:

- Intelligence and information sharing
- Screening, search, and detection
- Interdiction and disruption
- Forensics and attribution
- Planning
- Public information and warning
- Operational coordination

Many of these practices have clear implications and responsibilities within the realm of emergency management—particularly information sharing, planning, public information and warning, and operational coordination. (More information on each capability or function may be found in Chapter 9, Prevention Phase.) Also note that these capabilities are firmly established in the period before an incident actually takes place, rooted largely in the prevention, preparedness, and mitigation phases.

The National Response Framework (NRF) presents a National Strategy for Homeland Security in the United States, citing four key elements for DHS within this function:

- Prevent and disrupt terrorist attacks
- Protect the American people and our critical infrastructure and key resources
- Respond to and recover from incidents that do occur
- Continue to strengthen the foundation to ensure our long-term success

Again, emergency managers will see much that is familiar to them here. The elements cited straddle multiple phases of emergency management, and closely align with the NIMS model for incident management.

Becoming an active contributor within the homeland security function is thus becoming a more common requirement for an effective emergency manager in an urban environment. Ultimately, much of this capability comes from the emergency manager's liaison and communication function. As a position that often fills the void between more traditional functions such as law enforcement, fire services, and public works, emergency management offers an opportunity to "fill in the dots" that the various other agencies and functions may not observe.

Fusion centers also represent an important part of that effort in the United States. Established as a best practice by DHS, a fusion center acts as an information clearing house for public-safety agencies. There are approximately 70 such facilities across the nation, with at least one in all 50 states and in every major city. These centers routinely share intelligence and threat information with one another, and with participating agencies. Such a horizontally aligned system simply did not exist prior to 2001, and emergency managers are strongly encouraged to be sure that they are fully availed of the services and capabilities provided by their local center or centers.

Every major jurisdiction in the United States is also assigned at least one Protective Security Advisor (PSA). These DHS employees serve in the liaison function between the federal government and the private sector and are subject matter experts in securing critical infrastructure and key resources. As nearly 90% of the nation's critical infrastructure is either owned or operated by private entities, every urban emergency manager should be certain to be on a first-name basis this vital resource in his or her community.

The European Union (EU) has established an operations center similar in function to that of the American fusion centers. Called the EU Intelligence Analysis Centre, it is tasked with gathering and coordinating intelligence related to extremist activities on the continent. The center and its operations are led by an appointed Counter-Terrorism Coordinator and works closely with the transnational law-enforcement agency Europol.

## A Cautious Way Ahead

For emergency managers in urban centers, any honest assessment would indicate that terrorism is not going away any time soon. It will remain a threat for the foreseeable future, from both foreign terror organizations and domestic extremists.

A very real goal of terrorism is to goad governments and large organizations into overreacting—to perpetrate terror. Such an excessive response might come in the form of unrealistic and unsustainable spending or in imposing draconian laws or policies that undercut a jurisdiction's or corporation's core values.

As Jenkins and Godges note in the excellent *The Long Shadow of 09/11*, "America has always been an anxious nation, distrusting those in power, tending to see conspiracies where there are none, suspicious of newcomers, worried about disloyalty. There is in American society a deep-rooted tradition of political vigilance that at times borders on paranoia dating back to the 18th Century" (2011, p. 206). The same characteristic or trait is present in all established institutions within the democratic, developed world, be they a European parliamentary republic, or a major American corporation.

And so urban emergency managers will be asked to walk a very fine line in the coming years—to plan and mitigate for a relatively unpredictable threat but not to overcompensate at the expense of other risks or to sacrifice core values. In an era of tight budgets and sometimes competing objectives, this is far easier said than done.

## CASE STUDY: MUMBAI, 2008

The Islamic terrorist organization Lashkar-e-Taiba (LeT) was founded in approximately 1990 by self-proclaimed scholar Hafiz Muhammad Saeed. Around this time, the Pakistani and Indian conflict over the disputed district of Kashmir reached one of its most heated points of contention. LeT was envisioned as a defender of Islamic ideals in and around Pakistan and was soon supplying fighters to conflict zones in both Afghanistan and Kashmir. (The name refers to the "Army of the Good" or "Army of the Righteous," and variations include Lashkar-I-Tayyaba and Lashkar-e-Tayyiba.)

The group was clandestinely supported by Pakistan's intelligence branch, the Inter-Service Intelligence agency or ISI, as a proxy combatant with its perennial adversary India. Through the 1990s, the group conducted a number of attacks in and around the disputed Kashmir territories. But 2001 represented a transformation of LeT on a number of levels. Under pressure from the Pakistani government, the group changed its official name to Jamaat-ul-Dawa (also known as Jamaat-ud-Dawa), and professed to be an Islamic charity. Simultaneously in 2001, LeT—as it was still universally called—launched a series of brazen attacks on targets inside India. These included separate attacks on both the airport and a police station in Srinagar, and in apparent coordination with Jaish-e-Mohammed (JeM), an armed assault on the Indian Parliament in Delhi. The latter action nearly brought Pakistan and India into open conflict for a fifth time (Ghosh, 2010).

This tacit support was a source of great strife between the US, Indian, and Pakistani governments. Yet if an organization such as LeT, and its "front" groups such as Jamaat ul-Dawa, could continue to raise funds from the Pakistani diaspora and other private supporters even as it simultaneously received covert support from a national government, it represented a unique problem for other nations. As long as the government of Pakistan could maintain some degree of plausible deniability, it remained extremely difficult for the international community to impose the types of sanctions, trade penalties, and travel restrictions that often stall or reverse such behavior. (And the fact that Pakistan remains a primary supply route for the US and NATO war effort in Afghanistan was also of significant concern.)

And so, throughout the first decade of the new century, LeT operated multiple training camps with relative immunity. Some experts estimate that more than 150,000 militants were trained in these facilities, despite the fact that the general public had not yet heard of the group. This was soon to change.

In November of 2008, a well-documented assault on the Indian city of Mumbai left over 160 people dead, including several prominent police officials. Rarely in history have the details of a terrorist attack been so well captured in writing, on video, and via audio recordings. These records offer a unique opportunity to gain significant understanding of just how LeT operates, enabling this brief case study into the attacks.

The group had participated in a series of spectacular attacks on India over the previous 2 years, including a multiple bombing of passenger trains in Mumbai that killed 211 and injured over 1000. However, during the same period the group had also suffered losses in its leadership, including the killing of top lieutenant Abu Saad.

It was in this context that LeT undertook a series of even more spectacular attacks against their perceived enemies in India. According to testimony provided in US federal court, planning for this series of attacks began as early as the period from 2001 to 2004 (Hussain and Korecki, 2011). LeT was looking for a highly visible and disruptive attack on a major Indian city and had settled upon Mumbai. By 2008, a number of specific targets had been selected and detailed reconnaissance performed on each (Ghosh, 2010).

A group of around 20 young militants had been recruited and trained for the operation in India; ultimately 10 would be selected to actually participate in the suicide mission. The plan was quite complex and nuanced. A commercial vessel would be hired or hijacked to take the assault teams to the shores of Mumbai, where the ship's crew would then be killed. Once ashore, the LeT operatives would fan out and assault multiple high-profile targets across the city, creating as much mayhem and confusion as possible before being killed. It was, from the very beginning, to be strictly a one-way trip for the five two--man terrorist teams.

As is amply displayed in the media coverage that followed, this plan was highly successful. The terror cells infiltrated a number of sites, including a train station, a popular café, a Jewish center, and two luxury hotels. The groups used assault rifles and small explosives to attack concentrations of civilians and also fought pitched battles with responding security forces—who initially estimated the number of militants at 50–100.

The insightful HBO documentary *Terror in Mumbai* documents, in rare and detailed form, the uneven and disorganized response of local authorities. The running gun battles took 3 days to unfold, and essentially brought a major metropolis to a halt. Indelible images of the Taj Mahal hotel ablaze with the sound of gunfire echoing from within have led many, including political commentator Fareed Zakaria, to refer to the event as "India's 9/11" (Reed, 2009). When the smoke finally cleared, 166 victims were dead—primarily civilians—and scores more were injured.

Crucial to the initial investigation was the lone surviving gunman, Ajmal Amir Kasab. Stopped at a checkpoint by Indian security forces, Kasab and a companion tried to fight to the death, but the former was swarmed by police and taken into custody. In a chilling hospital interview, Kasab details the planning for the attack—clearly implicating both LeT and the Pakistani ISI (Reed, 2009).

Perhaps even more chillingly surreal were the audio recordings of the gunmen talking via cellular phone to their "controller" in Pakistan. Indian security forces managed to acquire the frequency being used by the terrorists as events were unfolding, and the resulting recordings offer incredible insight into LeT's operations. The young gunmen, anxious and sleep-deprived, are given their grim orders by a preternaturally calm advisor they refer to as Major Iqbal. If this "Major Iqbal"—which may or may not be his actual name or rank—is indeed a current or former member of the ISI or Pakistani military, it would indicate that a Pakistani official had full control of the terrorists on the ground in Mumbai as they murdered civilians (Hussain and Korecki, 2011).

Tellingly, from his hospital gurney after the attacks, Kasab stated that he was instructed to identify and execute Americans, Britons, and Israelis due to their perceived antipathy toward Muslims. Similar targeting has been seen in more recent attacks in Europe.

Following these attacks, ISI complicity with LeT operations was soon officially acknowledged in a US federal court. Admitted LeT plotter and US citizen David Headley assisted the group with reconnaissance and logistical planning for the 2008 Mumbai attacks. As part of a plea deal, he agreed to testify against an alleged financier and coconspirator for the attack, Tahawwur Rana. In May of 2011, in a US District courtroom in Chicago, Headley calmly recalled the close collaboration between the ISI and LeT in planning the attacks that killed 166 victims, including six Americans.

Headley, born as Daood Gilani in the United States but raised in Pakistan, said that he and Rana had helped plot the assault with input from both organizations. Under oath, Headley stated that "they coordinated with each other," and that "ISI provided assistance to Lashkar" (Hussain and Korecki, 2011). This crucial testimony, combined with the specific location of Al Qaeda leader Usama bin Laden at the time of his death only 3 weeks prior, certainly provides significant evidence that the ISI has been playing one side against the other on a grand scale. (The jury went on to deliver a split decision in its verdicts against Rana.)

In many ways, LeT represented a forerunner of extremist funding and operations—a prototype of a 21st century terrorist organization. It raised monies both clandestinely and openly through various channels, provided military-type training to a growing contingent of extremists, and pioneered the small-unit assault tactics that have been repeated and replicated in cities ranging from Paris to San Bernardino to Brussels. Emergency managers must be prepared for similar tactics to impact their organization or jurisdiction.

LeT has kept a low profile over the past several years, yet intelligence and counterterrorism experts state that the group is very much still intact and capable of planning another large-scale assault in the future. And it is not alone in that capability.

## SUMMARY

The professions of emergency management and homeland security are overlapping and intertwined areas of expertise, with multiple similarities and differences. Each is a rapidly expanding field in terms of academia and operations.

The threat of terrorism is a concern to both emergency-management and homeland-security practitioners and must be included in any complete risk assessment for an urban environment. Best practices dictate that emergency managers identify and assess types of extremism, and the operational tactics utilized by such groups. The terror attacks in Mumbai in November of 2008 illustrate the impact that such incidents can have upon a major urban center.

## References

Bevelacqua, A., & Stilp, R. (2009). *Terrorism handbook for operational responders*. Clifton Park, NY: Delmar.

Brill, S. (2016). Are we any safer? *The Atlantic, 318*(2).

Code of Federal Regulations. (1969). Washington, D.C.: U.S. Federal Government.

Department of Homeland Security (2002). *National strategy for homeland security*. Washington, D.C: U.S. Federal Government.

Ghosh, B. (2010). From Pakistan to the world. *Time, 175*(19). Retrieved from: <www.time.com>.

Haddow, G., Bullock, J., & Coppola, D. (2008). *Introduction to emergency management*. Burlington, MA: Butterworth-Heinemann.

Hussain, R., & Korecki, N. (2011). Terror case witness describes Pakistani intelligence-militant links. *Chicago Sun-Times*. Retrieved from: <www.suntimes.com>.

Jenkins, B. M., & Godges, J. P. (2011). *The long shadow of 09/11*. Santa Monica, CA: Rand Corporation.

Reed, D., (Director). (2009). Terror in Mumbai. *[Documentary]*. United States: Home Box Office.

Sylves, R. (2008). *Disaster policy & politics*. Washington, D.C: CQ Press.

## Further Reading

Ceniceros, R. (2004). Natural, manmade disasters require similar risk planning. *Business Insurance, 38*(32).

Department of Homeland Security (2013). *National prevention framework*. Washington, D.C: U.S. Federal Government.

CHAPTER 9

# Prevention Phase

## INTRODUCTION

Prevention may seem like a worthwhile yet elusive goal. There is the old maxim, "an ounce of prevention is worth a pound of cure." Sports teams trying to protect a lead late in a game or match are said to go into a "prevent defense." But how do we know if we actually prevented a detrimental event from taking place? How would an emergency manager know if he or she had successfully avoided a disaster?

After all, it is hard to prove a negative. If an emergency or disaster is prevented from ever taking place, how can we show empirical evidence of that fact?

Webster's defines the term as "the act or practice of stopping something bad from happening (Webster's dictionary (http://www.webster-dictionary.org/))." In terms of emergency management, prevention may be viewed as an action or series of actions that stops a disaster from taking place.

However defined, prevention is an emerging phase of emergency management in the United States. The Department of Homeland Security (DHS) and the Federal Emergency Management Agency (FEMA) have increasingly referenced prevention as being on equal footing and of equal importance to the other core phases or mission areas of emergency management—preparedness, mitigation, response, and recovery.

The five National Planning Frameworks are part of the National Preparedness System. Per FEMA, there is one Framework for each of the five preparedness mission areas, including prevention. Released in May of 2013, the National Prevention Framework compliments similar documents addressing the other phases of emergency management and acts in concert with the overarching National Preparedness Goal.



Urban Emergency Management. DOI: http://dx.doi.org/10.1016/B978-0-12-805307-2.00009-3

## ANALYSIS

With the possible exception of preparedness, prevention is the only phase that exists almost entirely *before* a disaster strikes and represents the one true opportunity to fully avoid a potential crisis. As is evidenced by the attention, it is receiving from the US government, prevention is emerging as the fifth pillar of emergency management.

### When Prevention Is Possible

It is first worth noting that prevention is not always possible. As noted in the Introduction, this phase of emergency management seeks to completely stop a disaster from occurring. This is a worthy goal, but it is not always attainable. Some guidelines are in order.

Quite simply, not all disasters can be prevented. Once a hurricane forms, no emergency manager can make it dissipate. When a line of severe thunderstorms drops several funnel clouds onto the ground, we cannot push them back up into the sky no matter what lies in their path.

Thus, it is important to note what types of hazards are subject to possible prevention. In reviewing the natural, human-caused, and hybrid types noted in Chapter 3, Natural Hazards and the Urban Environment, and Chapter 4, Human-Caused Hazards and Hybrid Hazards, several hazards seem difficult or impossible to prevent outright. This group would primarily consist of weather-related natural hazards beyond the current scope of human control, including tropical storms, tornadoes, tsunamis, and storm surges. Other hazard types appear more complex.

There is growing evidence that many small earthquakes are caused or at least influenced by human behavior such as fracking. However, larger earthquakes appear to also be beyond our ability to prevent. Some wildfires are natural and some are manmade, so this latter subset remains possible to prevent from threatening urban areas. Pandemics are often natural occurrences with human-causative factors contributing to their severity.

By contrast, the vast majority of human-caused and hybrid disasters are subject to prevention measures. Hazardous-materials releases, engineering failures, and cybercrime are all preventable in many cases. And perhaps the most preventable disaster type of all is terrorism. Given the prevalence of terrorist attacks within cities versus rural areas, the case for prevention to stand as a fully recognized phase of emergency management within urban settings grows stronger still.

By making these realistic recognitions early in the discussion—that not all hazards are fully preventable—the focus can be brought to bear on those which are. To summarize, our subset of potentially preventable hazards is then narrowed to:

- Cybercrime
- Engineering Failures
- Hazardous Material Releases
- Pandemics (partial)
- Terrorism
- Wildfires (partial)

Each of these potential disasters has unique qualities and aspects that grant the possibility of being prevented outright. Prevention would depend on effective assessment and action taking place prior to an actual incident. Fortunately, there is an established path to such actions, and it begins with identifying risk.

## Risk-Based Prevention

As is becoming a theme, an emergency manager's vital first step to prevention is to identify potential threats and hazards via risk assessment. To be capable of preventing a crisis or disaster, you must first know what can harm you (Woodbury, 2005). Identifying and ranking the greatest hazards to a particular organization or jurisdiction via a risk and hazard assessment is time well spent. A thorough assessment will be useful across all phases of emergency management. (Additional guidance may be found in Chapter 5, Assessing Risk in Urban Areas.)

The DHS program for antiterrorism-related prevention serves as an interesting point of reference. Following the recommendations of the so-called 9/11 Commission Report, a number of overarching objectives were identified:

- Building the homeland security enterprise
- Preventing terrorist travel and improving passenger screening
- Strengthening surface transportation security
- Strengthening global supply chain security
- Detecting and preventing biological, radiological, and nuclear threats
- Protecting critical infrastructure

While each objective could be very much viewed as a work-in-progress, such actionable high-level goals provide a worthy foundation for systemic change. In this case, the risk-oriented objectives are strongly aligned with the emergency-management phase of prevention.

This brings the conversation back once again to the goals and objectives identified by the National Preparedness System. Essentially, planning must take place to match available resources with potential hazards. Prevention may not be entirely possible, but even in such cases, assessment and planning efforts lead to mitigation of the identified hazards. As is the case with preparedness and mitigation, the prevention phase has no true beginning or end and should always be taking place at some level. Prevention planning will continue to be a critical element of any larger effort.

## Know What You Have

In conjunction with a risk or hazards assessment, it is highly beneficial for any emergency manager to inventory and analyze all critical infrastructure and key resources (CIKR) within his or her sphere of responsibility. Essentially, CIKR are assets essential to security, public health and safety, and economic vitality.

Writ large, this encompasses an immense range of equipment and facilities, including power stations, water-filtration plants, transportation systems, and government buildings. Yet, it would also include largely symbolic or iconic locations such as national monuments and cultural properties. Threats taken into consideration regarding such infrastructure must include both the physical and virtual or cyber realms.

In the United States, during the years following the September 11 attacks, the newly established DHS recognized that it simply did not have a complete picture of the systems that were vital to national security. An even greater revelation was the recognition that the vast majority of what could be considered CIKR was in fact a part of the private sector. Several programs were initiated to close this knowledge gap, including the vital National Infrastructure Protection Plan (NIPP).

The Critical Infrastructure Partnership Advisory Council (CIPAC) is one example of the federal government's efforts to gather more information about CIKR owned and operated by the private sector. Several sector-specific committees interact with their counterparts at DHS, providing two-way communication between these key stakeholders.

An even broader effort are the "datacalls" that DHS conducts with local, state, and private-sector partners. Subject-matter experts within a given jurisdiction, usually fusion-center or emergency-management representatives, are tasked with gathering and analyzing the CIKR within their area of operations. They then distill and repackage this information in a template provided by DHS and share their findings with that agency's Office of Infrastructure Protection.

By analyzing the assets and systems within a given geographical area, patterns of interdependence begin to emerge. The effort at the local level is time consuming, yet extremely enlightening. A vast web of interconnected nodes and subsystems are revealed, making the process extremely beneficial to any emergency manager able to take part.

A similar annual assessment is made of large-scale special events within a given state or urban area. The CIKR and special-event information then guides prevention, protection, and resiliency efforts at a national level and has a large impact on federal grant-funding amounts for large urban areas.

Updated by a presidential directive in 2013 (Haddow, Bullock, & Coppola, 2008), federal guidance on CIKR recognizes 16 (now 17) critical sectors that must be the focus of prevention and protection efforts:

- Chemical Sector
- Commercial Facilities Sector
- Communications Sector
- Critical Manufacturing Sector
- Dams Sector
- Defense Industrial Base Sector
- Emergency Services Sector
- Energy Sector
- Financial Services Sector
- Food and Agriculture Sector
- Government Facilities Sector
- Healthcare and Public Health Sector
- Information Technology Sector
- Nuclear Reactors, Materials, and Waste Sector
- Sector-Specific Agencies
- Transportation Systems Sector
- Water and Wastewater Systems Sector
- Election System (as of 2017)

This is unquestionably one of the areas where the line between emergency management and homeland security begins to blur. While emergency management follows an all-hazards model and homeland security is largely focused on the terror threat, in the phase of prevention, the two fields are virtually aligned. (See Chapter 8, Emergency Management and Homeland Security in Urban Areas, for additional information on the relationship between homeland security and emergency management.)

A truly effective prevention model would therefore not only look at risk or hazards but also focus resources on the core, critical elements of an organization or jurisdiction. Preventing the loss of or damage to CIKR assets represents a top priority for public-safety officials.

## The DHS Prevention Model

Released in 2013, the National Prevention Framework serves as the guiding document for disaster prevention by the US federal government. While the document focuses heavily on terrorism, its lessons certainly apply to other preventable hazards as well.

This guidance has evolved through a range of official documents, including Prevention Guidelines for Homeland Security and National Preparedness Guidance. Throughout this evolution, the prevention mission areas have focused on a handful of functions, including threat analysis, law enforcement, and information sharing. The National Prevention Framework introduces seven core capabilities required for the prevention of manmade disasters, each of which is worthy of further review.

### *Intelligence and Information Sharing*

The first core capability identified by the Framework is that of intelligence. In this case, intelligence refers to timely, accurate, and actionable information related to specific or imminent threats. From an emergency-management perspective, this would likely involve processing, analysis, and evaluation of information related to an imminent incident. This capability is based on the emergency manager's ability to interact with various entities in the public, private, and nonprofit sectors and to identify "outlier" information that may indicate a pending technological failure or act of terrorism.

### *Screening, Search, and Detection*

This capability relates to the ability to identify, discover, or locate imminent threats via observation and search. While clearly more focused on the law enforcement and private security sectors for the terrorist threat, such activities may involve emergency managers in the realm of engineering failures, pandemics, hazardous material releases, and wildfires. An alert emergency manager will recognize when conditions are ripe for any of these hazards to take place in his or her organization or jurisdiction.

### *Interdiction and Disruption*

From a terrorism standpoint, this capability related to diverting, intercepting, stopping, or apprehending the persons or materials involved in a potential attack. This particular law enforcement function is the traditional counterterrorism operation in which heavily armed (and armored) police units actively interdict a plot in progress. From an emergency-management perspective, it also translates to direct action—securing a hazardous material container before it ruptures, or reinforcing a failing bridge before it collapses. Whether in the homeland security or emergency-management context, interdiction translates to decisive action in the field.

### *Forensics and Attribution*

Forensics and attribution simply refers to using past materials to prevent future incidents. This could be the collection of physical evidence from a crime scene or the combination of such material with intelligence data. The goal is to learn from past acts to prevent future occurrences of the same type. Emergency managers would want to translate this skillset to engineering failures. For example, if a bridge failure was initially attributed to the type of gusset plate used in construction, all other bridges using the same type of plate should be examined to prevent a similar failure in the future.

### *Planning*

Planning in this context is related to developing crisis action plans that can be enacted on very short notice. In the realm of terrorism, it would allow for a number of preplanned options to be deployed upon the discovery of an imminent plot. In an emergency-management context, it would related to functional plans or annexes addressed to specific preventable hazards such as cybercrime or a manmade wildfire. In other words, preset plans can be put in place to be enacted the instant a given hazard is recognized as elevated or imminent.

### *Public Information and Warning*

This capability relates to the capacity to deliver coordinated, prompt, and reliable information to the public, or to particular subsets or sectors. Regardless of the threat or hazard, this capability remains vital to urban emergency managers. In a prevention context, such messaging and warning capability would allow stakeholders to be alerted to specific threats. Such a capability should be ideally viewed as a two-way path such as in the "See Something, Say Something" campaign.

### *Operational Coordination*

This capability refers to a Unified and Coordinated Command structure. Following the National Incident Management System (NIMS) model, operational coordination would maximize the unity of effort during periods of imminent threat. Such an operations-centric model would be applicable and useful should any preventable threat be perceived as growing in likelihood or potential to occur.

## Responsibility at All Levels

A core concept of the DHS National Prevention Framework, and of other preventative planning documents, is that of shared responsibility. No one agency or organization has the capacity to avoid every type of hazard listed as potentially preventable.

Individuals and communities can assist in preventing disasters by remaining observant of their neighborhoods, which they know best, and sharing any concerning developments with local authorities. This may include anything from concrete flaking off of a tunnel wall to suspicious chemicals being delivered to a vacant house. Such awareness may prevent a pending structural collapse or terror incident.

Nonprofits, nongovernmental organizations, and businesses share a similar expectation, but may also have additional, specialized abilities. These may include drawing on the expertise of academics within their ranks or helping to promote antiterrorism or antiviolence programs within their communities of influence.

Local governments provide services such as law enforcement, fire suppression, medical response, public works, and emergency management within their jurisdictions. Local governments also coordinate prevention efforts with neighboring jurisdictions and may operate fusion centers, or similar intelligence-sharing facilities.

State governments in the United States coordinate prevention activities in support of urban areas within their boundaries and operate fusion centers as a part of the larger national network. They also have specialized components—which may include law enforcement, forensics, engineering, and intelligence units—that could be particularly useful in many of the core capabilities related to prevention.

The National Prevention Framework also goes into some detail about determining imminent threats, or those which intelligence or operational information deems credible, specific, and impending enough to warrant implementation of additional measures to thwart any attack (Department of Homeland Security 2013, p. 3). Such information may be received via federal intelligence sources, via deployed law enforcement personnel or via private entities or the public.

Finally, the federal government has a wide range of capabilities related to the prevention of potential disasters. In the United States, the national intelligence community (IC), combined with the Department of Defense, Department of State, Federal Bureau of Investigation, DHS, and US Secret Service all have specialized units capable of interdicting or preventing large-scale terrorism or cybercrime incidents. The Department of Energy, Department of Transportation, and the US Corps of Engineers all have experts capable of preventing infrastructure failures. The Centers for Disease Control and Prevention and the Department of Health and Human Services have robust capabilities to detect and interdict in cases of potential pandemic.

## Prevention Resources

In Western nations, and in the United States in particular, politicians have set the expectation among the public that terrorism (and by extension, other manmade threats) can be prevented. A wide range of resources are available to emergency managers seeking to improve their program's stance on prevention. These include guiding documents and training options as well as increased public involvement.

### *Documentation and Training*

Prevention-related documentation and guidance is also becoming more commonplace and accessible. The National Prevention Framework summarized above provides a thorough review of prevention activities for terrorism-related incidents. FEMA provides this document, as well as the more comprehensive National Planning Frameworks that encompasses all five phases of emergency management, via its primary website at www.fema.gov. Even more broadly, the National Preparedness System seeks to apply an all-hazards model across the nation.

FEMA's Emergency Management Institute (EMI) provides other relevant courses to an emergency manager seeking to improve his or her prevention efforts. A number of online classes include strong, hazard-specific prevention elements, including "Workplace Violence Awareness" and "Critical Infrastructure Security: Theft and Diversion." And FEMA's National Training and Education Division lists no fewer than 66 courses in the Prevention mission area—more than in any other.

Not all training programs have to be as formal as an accredited college course or offered as part of a federal EMI session. Many law enforcement agencies and fusion centers have instituted their own internal training programs for terrorism prevention in particular, or have adopted and adapted federal courses to better fit their needs. One example would be the terrorism awareness and response academy (TARA), hosted by the Chicago Police Department, which contains particularly useful elements of interdiction and prevention. Emergency managers should not be shy about reaching out to local law-enforcement or fire-services resources in order to gain the necessary approvals to attend such vital training.

Drills and exercises will represent another crucial training piece. No corporation's or jurisdiction's emergency operations plan (EOP) can be deemed reliable or accurate until it has been thoroughly tested and vetted via drills and exercises.

Many such events will be centered around related functions such as public health or law enforcement. As an urban emergency manager should possess a broad general knowledge of overall crisis operations, such training is invaluable.

### *Fusion Centers*

Another vital resource available to emergency managers in large American cities is the network of fusion centers. A fusion center is essentially an interdepartmental intelligence and analysis hub located within a crucial jurisdiction. As of 2016, there were more than 70 fusion centers active in the United States, with at least one in every state capital and major metropolitan area.

Following the September 11 attacks, the IC, in particular, came under fire for having failed to "connect the dots" prior to the plot being carried out. A concerted effort was made to open effective channels of communication both horizontally and vertically across the IC, and among law-enforcement agencies. One of the most successful concepts was to essentially have some type of analysis field office out in the communities that these departments sought to protect. The idea of the fusion center was born.

In a sly play on the "if you've seen one, you've seen them all" cliché, intelligence personnel will state "if you've seen one fusion center, you've seen one fusion center." This is because while there is a general national standard of overarching purpose, the respective jurisdictions and host agencies are given wide latitude in the structure and format of their particular center.

From an emergency manager's perspective, his or her local fusion center likely provides intelligence and analysis on a range of threats not limited solely to homeland security. Agencies represented likely include local, state, and federal law-enforcement entities. Further skillsets may include fire services or public health. The vast majority of centers assess and analyze any threat to critical infrastructure or public gatherings, and this functional overlap make them an excellent potential emergency-management partner and resource in most urban areas. Ideally, this exchange can result in being not only a customer of the resulting briefings and alerts but also as a trusted subject-matter expert as well.

Emergency managers without an established relationship with their local fusion center may have to take the first step and provide an initial introduction. There will likely be a vetting or background clearance required to begin sharing information. Many European and Asian nations have similar programs and facilities, which may be beneficial to emergency managers in the public, private, or nonprofit sectors.

### *Public Eyes and Ears*

As noted previously, there is an important public information and warning component to a holistic prevention effort. This does not in any way infer that the general public should be made paranoid or coerced to spy on one another.

On the contrary, the civilian population is the master of their respective domains; they have an educated sense of what is normal during their daily commute, or an evening walk down their street. This group represents an immense, underutilized resource in the goal of preventing human-caused and hybrid emergencies. It is worth noting that public awareness is the first goal listed in the US DHS strategic plan.

The "See Something, Say Something" program is again worthy of mention. Introduced by the New York Metropolitan Transportation Authority (MTA), the program was adopted nationwide thanks in large part to its simplicity. The daily commuter knows if something is amiss at their usual station or on their usual train—they merely need to be empowered with how to effectively report such information.

Ultimately, prevention is simply not a burden or expectation that can be shouldered by the comparatively small emergency-management community. Civilian, academics, elected officials, and law enforcement partners will all play absolutely critical roles in disaster prevention in modern urban centers. A range of tactics, techniques, and procedures, which have been effective against the terrorism threat, have proven useful against other preventable hazards as well.

## CASE STUDY: CYBER INTRUSION

### The Facts

The beginning of the 21st century did not start well for international relations between Ukraine and Russia. The former Soviet republic went through a tumultuous series of leadership changes amid one corruption scandal after another. In 2014, pro-Russian president Viktor Yanukovych was ousted by street protests and replaced by a Western-leaning government. Russia reacted by publicly rejecting the new leadership, and soon initiated a covert invasion of the Crimean Peninsula, using "little green men"—Russian troops and equipment stripped of their insignia and unit markings. Eastern Ukrainian communities with large Russian ethnic minorities also revolted following prompting from Moscow. The tensions quickly devolved into armed conflict, which remains unresolved and ongoing years later.

Somewhat understandably lost in all of this high-profile international intrigue was the execution of the first confirmed cyberattack to succeed in damaging or destroying physical components of an active civilian utility. Just before Christmas of 2015, hackers based in Russia succeeded in shutting down a significant portion of the electrical grid in neighboring Ukraine. As a result, the Ukrainian incident becomes a case study in failed prevention.

International investigators believe that the attack was planned for many months, and included multiple probes into the targeted network prior to the actual disruption in December. Much of the illegal access was granted thanks to a lengthy "spear-phishing" campaign targeting managers and information technology employees for public works agencies in Ukraine. False but official-looking messages were sent to targeted individuals. If those officials clicked on a link embedded in the e-mail, a stealthy version of a program referred to as BlackEnergy3 was installed to run in the background of their machine. The malicious software, or malware, then allowed hackers to essentially log onto the administrative networks for the targeted utilities.

And yet this still did not allow them to reach their goal, as Ukrainian operators had placed a series of electronic security firewalls to divide the administrative systems from the Supervisory Control and Data Acquisition (SCADA) network that actually controlled the grid. But eventually, the hackers found their way onto the SCADA system as well and determined how to best disrupt the larger system. This thus becomes an object lesson in a missed prevention opportunity.

The actual attack first disabled back-up power systems, and then disabled the converters located at power substations. The converters original base programming, or firmware, was overwritten by a new malicious firmware that simply did not recognize commands coming from three regional control centers. Electrical power was summarily shut off to more than 200,000 civilians in the midst of winter, and Ukrainians public works managers initially had no idea how to switch it back on. (The affected breakers were eventually returned to manual operation on a site-by-site basis.) To add insult to injury, the hackers also flooded the utility's customer service center with bogus, computer-generated calls to effectively take it offline as well. Thus, began a new era of direct threats to SCADA systems vital to emergency managers in both corporate and public sectors.

## Analysis and Investigation

Could such a cyberattack occur in Western Europe or in the United States? In a word, yes. The same type of SCADA software running in Ukraine is operated in multiple other locations. In fact, *Wired* magazine noted that some investigators expressed that the firewalls and safety protocols present in Ukraine exceeded those of most US operators (Zetter, 2016). Given that members of the same Russian hacking unit made repeated attempts to interfere with the US election in 2016, the desire and intent for such a disruptive capability is highly probable.

The federal government is certainly taking the threat seriously. The National Infrastructure Protection Center (NIPC) was founded in 1998 to protect vital systems and has evolved significantly since. It is an interagency operations housed at the headquarters of the Federal Bureau of Investigation (FBI) and is one of many such efforts focusing on analysis, protection, and prevention. The NIPC focuses specifically on cybercrime, and its contributors include not just the FBI but also US intelligence agencies and the private sector.

In January of 2002, just months after the September 11 attacks, the agency distributed an alert that terrorists may seek to utilize spear-phishing attacks or similar network weaknesses to gain access to vital US utility systems. Shortly thereafter, in testimony before Congress, Special Agent in Charge Philip Thomas discussed the center's CIKR focus. "The National Infrastructure and Computer Intrusion Program also has a role in preventing terrorist acts. The focus of NIPC's 'Key Asset Initiative' includes asset identification and protection, and prevention and detection of computer intrusions. Assets include the major electrical, communications, and water facilities; transportation hubs; energy plants and other infrastructure which are instrumental in supporting societal activities and which, if attacked, would represent a major loss or disruption (Thomas, 2002)."

The NIPC has been accused of not effectively sharing information it receives. Yet that reputation appears to be improving, and the fact that the US federal government has recognized cyberattacks as a potential threat since at least 1998 is a somewhat encouraging sign. And in 2009, US Cyber Command was established, an armed forces command subordinate to Strategic Command. While focusing on national security issues, the effort unites various Department of Defense resources with similar goals and objectives to that of the NIPC.

A key objective of both Cyber Command and the NIPC is to interdict such attacks before they come to fruition. Preventing such disruptions is clearly far preferable than trying to cope with them once they have occurred. After all, winter temperatures in Chicago and Detroit are far colder than they are in western Ukraine.

### Case Study Conclusions

Preventing or interdicting malicious international actors is primarily the responsibility of the federal government. However, this responsibility seems to be shared across multiple departments in both the United States and in Europe. Such a lack of direct accountability often presages a failure to prevent a specific threat from occurring. Given the indisputable evidence that an international hack took down a significant chunk of the Ukrainian power grid in late 2015, US and European agencies clearly need to sharpen their focus on prevention efforts.

Emergency managers are not, in general, expected to be computer-systems experts. However, the Ukrainian episode proves that prevention is vital to the operation of critical infrastructure. It behooves modern emergency managers to be highly technically literate, as a failure to prevent such a cyber intrusion into their organization or jurisdiction can have very real impacts in the physical world around them.

## CONCLUSIONS AND IMPLICATIONS

The prevention phase represents the one true opportunity for emergency managers to avoid a major emergency or disaster. A brief listing of real-world prevention considerations for urban settings is warranted.

### Use Concentric Rings

No one phase of emergency management will provide all the solutions. As is evidenced by the Ukrainian cyber example, preventing a known hazard is not assured in all cases. Prevention should therefore be seen as a goal or objective, not as a freestanding solution to every risk or threat. Remember that emergency management is a cycle, and that each phase may overlap with others. A holistic emergency plan must be just that: whole.

### Know Your Infrastructure

A thorough hazards and risk assessment is a "must-have" for every emergency manager. It is also crucial to be a subject-matter expert on the critical infrastructure within your organization or jurisdiction. When allocating time and assets to protect vital operations, it is invaluable to know exactly which cogs are the most important to the function of the machine as a whole. Knowing what absolutely must be protected for the greater good brings an amazing amount of clarity to decision making while under pressure.

### Partner With Cops and Firefighters

There are traditional divides between public-safety agencies—police versus fire versus paramedics versus emergency managers. Those walls are a threat to public safety and should be gleefully breached. Do not take no for an answer and continue to make overtures to other departments until you find individuals with whom you can work. Local or state fusion centers are especially useful partners for emergency managers who have proven themselves to be reliable team players.

### Be a Resource for Others

Every urban emergency-management agency has a couple of "go-to guys." These are the men or women that spring to mind when others outside your agency or organization have information to share, or need information provided. Capable emergency managers can prove their value across multiple departments or lines of business, including life safety, risk management, resiliency, security, and logistics. Such a level of trust is especially crucial when seeking to prevent a crisis or disaster.

### Think Outside the Box

In a world of shrinking budgets and changing public perceptions, it pays to look for new ways of doing things. Use partnerships as a force multiplier by getting people with different skillsets to attack a challenge. Don't fear wearing many hats, including public speaker, grant writer, liaison, or logistics guru. Prevention is about getting out in front of a hazard, which likely requires new ways of looking at the problem.

## SUMMARY

The phase of prevention is emerging as the fifth pillar of emergency management. It offers the tantalizing and desirable possibility that many disasters can be prevented from ever occurring. A thorough assessment of risk and hazard should be accompanied by an analysis of CIKR.

In the United States, the Department of Homeland recognizes 17 critical sectors to protect and seven core capabilities to achieve the goal of prevention. A case study on the failure to prevent a cyberattack in Ukraine proves that prevention must be used in conjunction with the other phases of emergency management in order to guarantee the best likelihood for a positive outcome.

## References

Department of Homeland Security (2013). *National prevention framework.* Washington, D.C: U.S. Federal Government.

Haddow, G., Bullock, J., & Coppola, D. (2008). *Introduction to emergency management.* Burlington, MA: Butterworth-Heinemann.

Thomas, P. (2002). FBI Testimony before House Committee on Government Reform. Retrieved from: <https://www.fbi.gov/news/testimony/terrorism-preparedness-3>.

Webster's Dictionary. Prevention. Retrieved from: <http://www.webster-dictionary.org/>.

Woodbury, G. (2005). Measuring protection. *Homeland Security Affairs.* Retrieved from: <https://www.hsaj.org/articles/186>.

Zetter, K. (2016). *Inside the cunning, unprecedented hack of Ukraine's power grid. Wired.*

## Further Reading

Department of Homeland Security. (2011). Implementing 9/11 Commission recommendations. Retrieved from: <https://www.dhs.gov/implementing-911-commission-recommendations>.

Federal Emergency Management Administration. (2013). Retrieved from: <http://www.fema.gov/media-library-data/20130726-1914-25045-8890/hseep_apr13_.pdf>.

CHAPTER 10

# Preparedness Phase

## INTRODUCTION

Being prepared can mean different things to different people. Many of us may recall the motto to "always be prepared" from the Boy Scout activities of our youth. Perhaps you think of the famous quote attributed to Benjamin Franklin: "By failing to prepare, you are preparing to fail." Or maybe it recalls the words of famous inventor (and stickler for detail) Alexander Graham Bell: "Before anything else, preparation is the key to success."

Whatever image or quotation comes to mind, preparedness is unquestionably a valuable commodity. Webster's defines the term as "the fact of being ready for something." Thus, preparedness is inexorably linked to some future event or incident. And in the realm of emergency management, preparedness may be defined as a state of readiness to respond to a disaster, crisis, or any other type of emergency situation (Haddow, Bullock, & Coppola, 2008, p. 183).

However defined, preparedness is a cornerstone of emergency management. The preparedness cycle essentially involves four core principles: assessment, planning, preparation, and evaluation. All of which leads to a core question for students or practitioners in the field: What does real preparedness look like, and how do I get there?

## ANALYSIS

Generally acknowledged as one of the first phases of the emergency-management cycle, preparedness almost inarguably does not get the attention it deserves. Alongside prevention, preparedness is the only phase that exists almost entirely *before* a disaster strikes, and represents one of the few true opportunities for an organization to ready itself for a potential crisis. Haddow's definition of preparedness involving a state of readiness



Urban Emergency Management. DOI: http://dx.doi.org/10.1016/B978-0-12-805307-2.00010-X

is again useful. (Haddow et al., 2008, p. 183). In fact, the terms preparedness and readiness are often used interchangeably.

This is a key insight into the phase of preparedness; it represents an organization's primary opportunity to ready itself for tough times. Mitroff concurs, noting that "an organization's major business objectives are less likely to be derailed" when the group has undertaken basic preparedness efforts (2001, p. 29).

But what is true preparedness? How does an organization or jurisdiction begin preparing? The answer, in large part, depends on the threats and hazards faced by an organization. A business could be faced with product recalls, industrial accidents, boycotts, hostile takeover attempts, or strike threats. A governmental jurisdiction could be faced with natural disasters, civil unrest, or hazardous materials releases, for example.

## Start at the Beginning

The key first step is therefore to identify potential threats and hazards via risk assessment. This term refers to the process or methodology of evaluating risk (Haddow et al., 2008, p. 68). In essence, to be prepared for a crisis or disaster, you must know what can harm you. Preparation therefore begins with identifying and ranking the greatest hazards to a particular organization. These may be natural or manmade and will vary greatly with geography, business type, size, etc. Results and outcomes will be highly dependent upon this thorough analysis. (Detailed analysis may be found in Chapter 5, Assessing Risk in Urban Areas.)

After identifying the most likely or harmful threats or hazards, an organization must then begin countering potential impacts. This is the portion of preparedness that involves making every internal effort to prevent or counteract given threats and hazards. Essentially, planning must take place to match available resources with potential hazards. And that process leads directly into the next phases of emergency and crisis management.

One of the key factors to an effective response is what transpires previously. Haddow et al. define preparedness as "a state of readiness to respond to a disaster, crisis or any other type of emergency situation" (Haddow et al., 2008, p. 183) and mitigation as "a sustained action to reduce or eliminate risk to people and property from hazards and their effects" (2008, p. 75). These two phases must thus be well underway prior to a large-scale incident occurring.

This brings the focus back once again to the goals and objectives identified by the National Response System/Framework. Any comprehensive and holistic emergency-management program must include a detailed risk and hazard analysis that directly informs mitigation and preparedness efforts. In many ways, the process of risk and vulnerability assessment may be considered endemic to preparedness. Assessment represents the first core principle within the cycle of preparedness. With potential threats and hazards identified, strategies can be developed for their minimization or elimination.

## Make a Plan, and More

The clarion call for many local, state, tribal, and federal preparedness programs in the United States has been some variant of "make a plan, build a kit, get information." This simple assessment provides good insight into the next step in preparing for a crisis or emergency.

Whether making preparations for an individual, a family, an organization, or a jurisdiction, an emergency plan is a critical goal. For an individual, this first step may be as simple as a one-page list of emergency contact numbers, medications, and meeting locations that can easily be carried in a wallet. For an organization or jurisdiction, it will require considerably more effort—hopefully resulting in a holistic emergency operations plan (EOP). As an emergency manager, our role is to constantly push this planning effort closer to currency and completion. Planning is the second core principle of preparedness.

The next step again reveals a continuum between individual and organizational preparation. A "kit" for an individual may include extremely basic items such as food and water, medication, a first-aid kit, walking shoes, and a change of clothes. Again, jurisdictions and organizations are using a scaled-up version of the same concept. The risk and vulnerability assessment and emergency plan will inform decisions at this level, based on a few key questions.

What vulnerabilities represent a real risk to normal operations? What redundancies or materials would reduce or eliminate that interruption? How and where should those back-up supplies or processes be put in place?

Honestly answer these questions and take action based on the results. Preparation is the third core principle of preparedness. It is also, without question, the most labor-intensive and costly. True preparation may require altering existing processes or procedures within an organization, and such change will always be resisted by internal mechanisms. It may also involve stockpiling supplies, or purchasing new equipment or software that will be useful in a time of crisis. Such expenditures may also promote resistance. And preparation will always require additional training, drills, and/or exercises to understand and effectively utilize the newly applied procedures and equipment.

Evaluation is the fourth and final core principle of preparedness. Once an assessment has been performed, a plan has been developed, and preparations have been made, the result must be honestly reviewed. Again, for an individual this might involve simply taking stock of his or her preparedness efforts and determining what needs improvement. For an organization or jurisdiction, it should include a thorough review and reassessment of all threat and vulnerability assessments, any resulting planning or equipment enhancements, and any training or personnel programs.

Preparedness can in many ways be viewed as the first step in any emergency-management cycle. It should also be a constant in all well-managed organizations and jurisdictions. Even in the midst of the response, recovery, or mitigation phases of an event, a good emergency manager will have one eye toward preparing for the next crisis.

Preparedness is also often a highly cost-effective function within emergency management. Along with mitigation and prevention, preparedness can have a significant impact *before* a given crisis erupts. Ultimately, preparedness is directly related to the efficiency of the planning for and response to a given hazard.

Frequently, such hazards are fairly obvious and can be addressed directly. As noted by Fischer and Green, "threats may be relatively predictable for a given site, such as the possibility of explosion, a spill, or a leak in a chemical plant" (1998, p. 293).

However, not all hazards are so obvious—which is why the preparedness cycle is such a crucial and useful tool for emergency managers. The cycle includes the overarching functions consisting of assessment, planning, preparation, and evaluation. Yet at its core are a group of direct actions which can lead to greater overall preparedness. These include assessing the threat, assessing the vulnerability, identifying shortfalls and requirements, implementing enhancements, exercising and training, and reassessing progress.

It is these concrete actions that make the preparedness cycle much more than an academic exercise. It is in fact a highly useful and practical tool for emergency managers to improve their response to a wide range of threats and hazards.

Multiple professional organizations and planning documents exist which can be of great use to an emergency manager. There is often no need to reinvent the wheel. One example is the National Higher Education Emergency Management Program Needs Assessment released in late 2016 by the

National Center for Campus Public Safety. Based on an extensive survey of college safety professionals, the assessment includes findings and recommendations for the higher education sector, as well as a listing of possible resources and relevant organizations. It is just one sample from one sector in a continuum of many such available studies and guidance. (Further information on cross-sector collaboration may be found in Chapter 7, Collaboration in Urban Environments.)

As with any cycle, preparedness has no true beginning or end. Once a truthful review of previous efforts has been performed, an effective emergency manager dives right back into assessing threats and vulnerabilities for his or her organization. To review, the core tenants of preparedness are:

- Identification of threats and hazards
- Planning to counteract risks
- Preparation of planned actions
- Evaluation of process

## Serving the Whole Community

As the Federal Emergency Management Agency (FEMA) proceeded through its own preparedness process in the first part of the new millennium, it recognized a shortcoming. A large segment of the American population could not be lumped into a generic or standardized group and given the same treatment before, during, and after a disaster. Residents were not homogeneous and could not realistically be treated as one population with equal capabilities.

This fact became all too clear in the disastrous failure of the US emergency-management system as a whole during Hurricane Katrina in 2005. (For more information, see the case study in Chapter 2, Development and History of Emergency Management.) The vast majority of deaths along the Gulf Coast were disadvantaged in some serious and identifiable manner, which directly contributed to higher fatality rates among these subgroups.

Emergency planners on the federal, state, and local level recognized that a "one size fits all" emergency plan was a failure waiting to happen. Extensive effort was put into identifying and understanding at-risk populations and determining how to best incorporate such subgroups into the overall emergency planning picture. FEMA dubbed the effort "whole community."

As Haddow et al. note, no set criteria automatically demarcates what does or does not constitute a special-needs population (Haddow et al., 2008, p. 192). Each community bears some responsibility in identifying those members who may require additional or accelerated assistance in times of true crisis. A partial list would include:

- Children
- The elderly
- The chronically ill or hospitalized
- The disabled
- Immigrant populations
- Non-English-speaking populations
- Minority populations
- The poor or homeless
- The illiterate
- The mentally ill
- Prisoners

Additional groups may include the transitory or those unfamiliar with a given area such as tourists staying in hotels or college students in dormitories. Planners soon recognized that the key for whole-community efforts to succeed was preparation and outreach. Each of these groups would be starting at a significant disadvantage if they were reached when a disaster had already occurred—a more proactive approach was required.

Precisely because of its preevent nature, preparedness is the ideal phase of emergency management at which to interdict on behalf of individuals with special needs. There is no clear and easy path to addressing these myriad challenges. It must be done on a case-by-case basis prior to any crisis. Yet several broad strategies are likely to be effective in advancing the whole community planning approach, including:

- Availability of key materials in graphic form and in multiple languages
- Advance registry of individuals by need and location
- Effective mass messaging systems
- Disability training for first responders and volunteers
- Specific and tested transportation means for individuals with physical disabilities
- Active partnerships with hotel and dormitory management
- Special transportation and shelter for incarcerated evacuees
- Active partnerships with disability experts and advocates
- Special shelter facilities and caregivers for young or mentally disabled evacuees
- Availability of key assistive devices and common medications at shelter locations

While this list is not all-inclusive, it provides emergency managers valid starting points to address several common challenges. Once again, it is critical that managers form strong partnerships with subject-matter experts in the nonprofit and volunteer communities during the preparedness phase. Their knowledge and assistance will be absolutely crucial to the special-needs population during periods of crisis.

## Should I Stay or Should I Go?

One of the classic and most common quandaries faced by emergency managers, particularly those in an urban setting, is whether to issue evacuation or shelter-in-place orders. While the actual evacuation or sheltering will take place during the mitigation or response phases, it is critical that planning and education take place prior to a large-scale incident.

Concentrated civilian populations present unique challenges to emergency managers. Legal and legislative rules vary from state to state and region to region regarding the authority of a chief elected official, or the head of a public-safety agency, to order a mandatory evacuation. Whether mandatory or recommended, studies have shown that up to 20% of the population will ignore or willfully disobey such guidance. While evacuation or sheltering are two crucial tools for public-safety officials, the fact is that people are stubborn and sometimes fail to take actions in their own best self-interests.

And yet, despite these challenges, evacuation and shelter in place remain two of the most powerful weapons in an emergency manager's arsenal. Such plans allow authorities to either move a population out of harm's way or to have them hunker down until the most serious danger passes. Advance planning is absolutely vital, as is notification and public education.

An urban emergency manager's first step should, as always, be a thorough and truthful risk assessment. The next step in the planning process is to put likely risks and hazards into the evacuation column or the shelter column. While all emergencies are unique, their commonalities allow most dangers to be so predesignated.

Shelter is generally required for no-notice severe-weather events such as tornadoes, or hazardous materials releases where outdoor air quality is an immediate health risk. Sheltering essentially allows the hazard to pass, or to be mitigated or neutralized by responders, while civilians remain in a relatively safe and secure location.

Incidents that provide some advance notification, such as hurricanes and tsunamis, would necessitate evacuating civilian populations. The second common type of evacuation driver is damaged infrastructure such as a gas leak or structural damage from an earthquake. Essentially, citizens should be moved only if they are in immediate danger. Every emergency manager should have a short list of what hazards his or her organization or jurisdiction faces, and whether evacuation or sheltering in place is the most likely recommended course of action for each.

It should also be noted that the extent of such authority is on a state by state, or even jurisdiction by jurisdiction, basis in the United States. In some locales, the mayor or governor is empowered to issue mandatory evacuation orders during a state of emergency or to order all nongovernmental vehicles off of the roadways. In other jurisdictions, this authority is limited to guidance or recommendation only. It is imperative that such authorities be listed in detail within the current EOP, and that senior elected and public-safety officials be briefed to fully understand the extent or limitation of their legal powers.

Transportation options will be critical in cases of evacuation or relocation, especially for populations without access to their own vehicles. Explicit agreements, such as signed contracts or notarized memoranda of understanding, should be initiated for multiple transportation modes, including buses or trains where available. Recall that any effective plan must account for moving a very large number of people in a very short amount of time, and all potential routes should be considered.

Some special populations, such as nonambulatory hospital patients, may simply not be able to evacuate and will be required to shelter in place. A thorough plan should take such points of refuge into account and make them a top priority during the response phase of an emergency.

## Resiliency

Another form of preparedness is resiliency, which refers to making a place or thing more resistant to damage or disruption. This could mean strengthening a specific resource against a specific threat, sometimes called hardening. It could also refer to building in systemic redundancies to minimize the uniqueness or vital nature of a single asset or node.

On the public-sector side, such efforts are generally referred to as continuity of government (COG) or continuity of operations (COOP). Following the events of September 11 in particular, American governmental agencies have taken a much larger interest in critical infrastructure and services (Haddow et al., 2008, p. 208).

The private sector has also embraced this concept to a large extent, in the form of business continuity efforts. In essence, business continuity planning refers to documenting the processes and procedures necessary to keep the company or organization running during and after a crisis. It often involves redundant data storage, alternative work sites, and cross-training personnel. The ultimate goal of a business-continuity or business-resilience plan is for zero recovery time. In other words, even when faced with a severe hazard, failover systems immediately kick into place, eliminating any period of significant disruption to the organization.

Multiple professional organizations exist catering to experts in the field, including the Association of Contingency Planners (ACP), the Business Continuity Planners Association (BCPA), and the Business Resumption Planners Association (BRPA). The private sector remains largely ahead of the public sector in such planning, and emergency managers are strongly advised to avail themselves of their fellow professionals.

While such resilience and continuity efforts certainly begin in the preparedness phase of emergency management, note that they will prove critically important during the mitigation, response, and recovery phases as well.

## Training and Education

One of the core applications of preparedness is training and education. This process applies not only to emergency managers and their fellow responders but also to the public. A public-safety official's best possible ally is an educated and prepared populace.

To the shock of not a single reader, such programs build upon an organization or jurisdiction's risk assessment and EOP. Once completed, these will serve as guiding documents for an effective training and education program. Such a program should focus on the hazards and risks faced by its parent entity.

FEMA's Emergency Management Institute (EMI) provides an excellent point of reference. Located on a former college campus on the outskirts of Emmetsburg, Maryland, EMI is one of the finest public-safety training facilities in the world. Offering 100 courses in both online and in-residence formats, the institute has trained literally thousands of responders, officials, and volunteers. Additionally, certified EMI instructors fan out to state, tribal, and local jurisdictions across the nation to present specialized training.

And while the focus is placed squarely on emergency management and the National Incident Management System (NIMS), participants and students include a range of professionals including elected officials, and members of the National Guard, fire services, and law enforcement. The courses also include hazard-specific topics, offering real-world lessons on hurricanes, tornadoes, and earthquakes. Given the depth and breadth of its offerings, emergency managers should unquestionably consider EMI to be a primary training and education resource.

As the relatively young field of emergency management becomes more broadly accepted and proves its utility, the number of college-level courses has grown exponentially. In fact, EMI has partnered with multiple institutions via its Higher Education Project (Haddow et al., 2008, p. 196).

From just a handful of programs or certificate offerings at the beginning of the century, *Emergency Management* magazine now lists at least 70 accredited high-education institutions offering approximately 300 degrees. That number is growing every semester, as additional programs, departments, degrees, and certificates are added. Students and practitioners of emergency and crisis management now have formalized education opportunities that simply did not exist a decade ago.

Yet not all training programs have to rise to the complexity of a top-level EMI course, or to that of a university-level curriculum. Multiple jurisdictions have instituted their own internal training programs for responders and volunteers, or have adopted and adapted federal courses to better fit their needs. There are as many variables as there are emergency managers.

One example of this training is the Community Emergency Response Team (CERT) program. CERT members are community volunteers who receive free hands-on and classroom training on a number of basic emergency-management topics, including first aid, light search and rescue, and basic firefighting. Graduates of the program are then assigned to geographic teams that serve as "second responders" when there is a major emergency. It has often been the case in times of crisis that all police officers, firefighters, and emergency medical technicians are occupied with the immediate emergency and unavailable to provide additional services. In Europe, the European Voluntary Service/Erasmus + program offers some roughly equivalent opportunities. (For more information on CERT, see the brief case study at the conclusion of this chapter.)

### Drills and Exercises

Another crucial element to the overall training piece is the concept of drilling and exercising existing plans. No EOP can be deemed reliable or accurate unless it has been poked, prodded, and tested via drills and exercises.

Such efforts can be largely divided into two categories: discussion-based exercises and operations-based exercises. Discussion-centric exercises include seminars and workshops as well as traditional tabletop exercises. Operations-centric exercises include drills and full-scale exercises {{{Fig. 10.1}}}.

Most large cities in the Unites States subscribe to the qualifications and principles of the Homeland Security Exercise and Evaluation Program (HSEEP). Overseen by FEMA, HSEEP "provides a set of guiding principles for exercise programs as well as a common approach to exercise program management, design and development, conduct, evaluation, and improvement planning."

**FIGURE 10.1**
A full-scale exercise at Midway Airport in Chicago includes fire services personnel as well as volunteers from the US Navy and the local CERT chapter.

HSEEP essentially provides guidelines for any emergency-management or homeland-security drill and exercise program by normalizing the training cycle into four segments: design and development, conduct, evaluation, and improvement planning. To be HSEEP compliant, a program must successfully meet all four criteria on a consistent basis. HSEEP also drives an organization or jurisdiction to run interrelated drills and exercises, starting with a simple discussion-based event and escalating to a full-scale exercise. While the former may be as simple as gathering a few key players around a conference table, the latter will involve responders and equipment in the field dealing with a simulated large-scale emergency.

### *Professional Accreditation*

One sure sign of the field of emergency-management's growing acceptance is the notion of professional accreditation. In addition to academic degree programs, endorsements by national and international professional organizations have now been established. (This receives further discussion in Chapter 8, Emergency Management and Homeland Security in Urban Areas.)

Individual emergency managers now have a number of options for professional certification. Multiple states within the United States now have certifications or accreditations overseen by the state emergency-management agency or by the state university system.

The most widely recognized board certification within the profession remains the Certified Emergency Manager (CEM) credential from the International Association of Emergency Managers (IAEM). To be eligible, individuals must meet certain requirements including several years in the profession, 200 hours of formal emergency-management and general-management training, and the completion of a comprehensive essay exam.

Overall institutions and departments also have accreditation options. The Emergency Management Accreditation Program, or EMAP, was founded in the early 2000s by a dozen organizations and agencies, including the aforementioned FEMA and IAEM, and the National Emergency Management Association (NEMA). Established essentially as a nonprofit company overseen by these three organizations, EMAP seeks to standardize the field by a system of accreditation and peer review, with a stated vision of "safer communities through measurable standards of excellence for emergency-management programs." Local, regional, tribal, and state programs are eligible to apply for the voluntary process.

Participation in such peer-reviewed efforts by individual emergency managers, or by their departments as a whole, should be considered a highly desirable portion of the training and education element of preparedness.

## Public Education and Notification

Finally, no preparedness program would be complete without a healthy public-education component. This does not mean that the general public must become disaster experts; such a goal is likely unattainable. Rather, the civilian population needs to have a good basic awareness of potential hazards and have some idea of how they will receive or retrieve additional information. In fact, public awareness is the very first goal listed in the US Department of Homeland Security (DHS) strategic plan.

The "See Something, Say Something" campaign is a very strong example of a successful public-awareness effort. Originally implemented by New York's Metropolitan Transit Authority (MTA), the core messaging has been generally adopted nationwide by DHS and by multiple transit agencies in particular. S4, as it is known in public information officer shorthand, checks several key boxes for such programs. It is simple and actionable, empowering public involvement on matters of crime or terrorism via a defined action. It thus involves the public—particularly transit commuters—as force multipliers in the form of additional eyes and ears to assist public-safety officials in both the prevention and preparedness phases. (And the alliteration of the catchphrase doesn't hurt either.)

The S4 effort may be seen as a "pull" campaign, as it seeks to gather information, in this case regarding potential suspicious activity or items, from the public. A "push" campaign sends information the other way, usually in the form of media messaging or a mass notification or alert. Virtually every large American city has a mass-messaging platform for mobile-device users, Notify NYC and NotifyChicago being two notable examples. Using cellular and Wi-Fi technology, these notification systems allow public-safety agencies to push out information in ways that simply would not have been possible a decade ago. End users may tailor their accounts to determine what types of alerts they desire and receive the service for no cost beyond their typical data plans. Of course, emergency notifications will continue to use more traditional channels such as television, radio, and news websites.

From a preparedness standpoint, a basic public awareness combined with the ability to reach large swaths of the general population during a crisis represents a powerful one-two punch. The ultimate goal is to switch seamlessly and rapidly from the largely preincident preparedness cycle to the actionable response cycle.

## CASE STUDY: THE COMMUNITY EMERGENCY RESPONSE TEAM

### Overview

As previously noted, the CERT is a concept designed to train second-line responders available to assist themselves and others during times of crisis. CERT seeks to fill a gap by providing services ranging from communications and transportation to traffic control and medical triage. Readily identifiable by their bright green helmets, CERT members have been deployed hundreds of times across the United States in recent years by public-safety agencies.

First implemented in California in the mid-1980s during a period of multiple significant earthquakes impacting the state, the concept started by the Los Angeles City Fire Department was soon adopted by multiple neighboring jurisdictions. Los Angeles had in turn based its program on lessons learned from civilian volunteers in Japan, particularly following the Kyoto earthquake of the same period. CERT is now a nationwide program, overseen by FEMA since 1993, and with 2600 chapters established in virtually all medium and large American cities, it is a concept which has certainly caught on.

The CERT website lays out a series of objectives for the program. First, present citizens the facts about what to expect following a major disaster in terms of immediate services. Second, give the message about their responsibility for mitigation and preparedness. Third, train them in needed life-saving skills with emphasis on decision-making skills, rescuer safety, and doing the greatest good for the greatest number. Fourth, organize teams so that they are an extension of first responder services offering immediate help to victims until professional services arrive.

Ultimately, the effort seeks to organize and train civilians who would likely become spontaneous volunteers in the event of a disaster or catastrophe affecting their city. A frequently used slogan within the training program is "doing the greatest good for the greatest number of people." It may thus be seen as a force multiplier for local and regional response agencies. A 2012 survey by the Los Angeles CERT organization indicates that 90% of CERT chapters offer their initial training at no cost to participants, thus removing a serious potential barrier of entry for some.

### In Practice

The program closely aligns with the core tenants and characteristics of the preparedness phase—a state of readiness to respond. CERT chapters may vary from one jurisdiction to another, but all have the same basic, introductory training and orientation requirements, with seven core components:

- Disaster preparedness
- Disaster fire suppression
- Disaster medical operations (First aid)
- Light search and rescue operations
- Disaster psychology
- Team organization/management
- Program review and live simulation

Once this core curriculum has been completed, volunteers are free to pursue further emergency management and preparedness training and education. Many chapters utilize both internal experts and external speakers to further their preparedness skillsets. Some even have additional highly defined levels or tiers of training and qualifications.

Field deployments are largely coordinated via a sponsoring response agency such as the local police or fire department. Some CERT chapters are aligned directly with FEMA or with the local emergency-management agency for a given jurisdiction. CERT teams have been deployed in urban environments following earthquakes, fires, and flooding. They have also been utilized as volunteers in sheltering operations and in large-scale special events, drills, and exercises. And many states and regions also sponsor skills competitions among chapters.

Given the limited emergency services budgets of all large cities, citizen-activist programs such as CERT make operational and fiscal sense. Moving forward, skilled emergency managers are likely to increase their coordination with such trained volunteer organizations.

## CONCLUSIONS AND IMPLICATIONS

The preparedness cycle represents one of the few, true opportunities for emergency managers to intervene prior to a major emergency. A brief listing of real-world preparedness considerations for urban settings is warranted.

## Information is Power

Preparing for a disaster really involves two key components—finding out what you have and determining how to protect it. Again, this requires an effective and honest risk and hazard assessment to determine potential dangers. With that knowledge in hand, it is key to determine how to prepare for specific hazards and how to best protect critical infrastructure and populations. It all starts with knowing what can hurt you and how.

## The Cycle Works, So Use It

The preparedness cycle involves four key processes: assessment, planning, preparation, and evaluation. Any emergency-management program in an urban organization or jurisdiction is currently in one or more phases of this subcycle. Motion is the key. A good risk assessment and EOP should be validated and tested, and then the process should start anew. True preparedness means not resting on your laurels.

## The 80/20 Rule

It is frequently said that 20% of your population will require 80% of your time and resources. Individuals with disabilities or other special needs may require outsize effort during crises. The time to come to that realization is prior to a major emergency, not during. It is absolutely critical that a successful emergency manager coordinate sustained and effective outreach to these populations under his or her watch.

## Harden the Target

"Hardening the target" is a phrase long used by the military and security professionals. This essentially refers to making an adversary's job tougher. Whether the hazards facing your organization are natural, human-caused, or hybrid in nature, this is a crucial element of preparedness. Resiliency, redundancy, and continuity are the enemies of chaos and are thus the best allies an emergency manager can have.

## Words Matter

Strong public education and emergency notification programs will prove invaluable during an actual emergency. If you are starting a crisis with a populace that possesses a good, basic understanding of public safety, and you have a methodology to get real-time information to the group, half the battle has already been fought and won. It is absolutely worth emergency managers putting in such effort on the front end, because the alternative is not pretty.

## SUMMARY

Preparedness represents a rare opportunity for an emergency manager to get out in front of an emergency or disaster. Risk assessment and emergency planning allow effective preparations to be focused and targeted. Different portions of the population will have different needs, particularly in crises involving large-scale evacuation or shelter-in-place efforts.

Emergency management has evolved substantially in recent years with exponential advances in professional training and education opportunities. Resiliency programs and an increased focus on educating and alerting the public prior to a major incident are efforts likely to pay large dividends during a disaster. Mass messaging and real-time communication will be crucial as a hazard shifts from a possibility to an actuality. Finally, the CERT program is an example of a preparedness-oriented volunteer effort.

## References

Fischer, R., & Green, G. (1998). *Introduction to security*. Burlington, MA: Butterworth-Heinemann.

Haddow, G., Bullock, J., & Coppola, D. (2008). *Introduction to emergency management*. Burlington, MA: Butterworth-Heinemann.

Mitroff, I. (2001). *Managing crises before they happen*. New York, NY: AMACOM.

## Further Reading

Emergency Management Accreditation Program. (2016). Retrieved from: <https://www.emap.org/>.

Emergency Management. (2016). Retrieved from: <http://www.emergencymgmt.com/degrees>.

Federal Emergency Management Agency. (2013). Retrieved from: <http://www.fema.gov/media-library-data/20130726-1914-25045-8890/hseep_apr13_.pdf>.

Federal Emergency Management Agency. (2016). Retrieved from: <https://www.fema.gov/about-community-emergency-response-team>.

International Association of Emergency Managers. (2016). Retrieved from: <http://www.iaem.com/>.

Los Angeles Community Emergency Response Team. (2016). Retrieved from: <www.cert-la.com>.

National Center for Campus Public Safety (2016). *National higher education emergency management program needs assessment*. Eugene, OR: University of Oregon.

Webster's Dictionary. Preparedness. Retrieved from: <http://www.webster-dictionary.org/>.

CHAPTER 11

# Mitigation Phase

## INTRODUCTION

The phrase mitigate can hold very different meanings in different contexts. Webster's defines it as "to cause to become less harsh or hostile: mollify." The Federal Emergency Management Agency (FEMA) defines mitigation as "those actions taken to reduce the exposure to and impact of an attack or disaster" (2003, p. B25). Roberts identifies the goal of mitigation to "decrease the damage disasters cause to lives and property by anticipating hazards and making communities more resistant to their effects" (2009, p. 127).

Mitigation can thus apply to natural, human-caused, or hybrid hazards. Multiple strategies and programs exist at the local, state, regional, and federal level to accomplish such an objective. In many ways, mitigation is the most long-term and far-reaching phase of emergency management. Improving levee design and relocating flood-prone structures, for example, could have a positive, multigenerational impact on a community.

Mitigation also requires a different skillset than the operational phases of emergency management such as response and recovery. Haddow, Bullock, and Coppola note that truly successful mitigation manager must possess political intuition and planning expertise as well as public-relations and consensus-building skills (2008, p. 76). Mitigation is thus both overarching within emergency management and demands a very particular skillset. Somewhat due to this broad nature, the political, corporate, and budgetary pendulum sometimes swings in favor of mitigation's all-hazards applicability, and sometimes does not.

## HISTORY AND CURRENT CONTEXT

Somewhat uniquely, mitigation can be actively underway before, during, and after a major incident. Mitigation can come in the form of a new seawall or



Urban Emergency Management. DOI: http://dx.doi.org/10.1016/B978-0-12-805307-2.00011-1

fenceline put in place before a natural disaster or terrorist attack. Yet it can also come in the form of tactics utilized by firefighters or local emergency managers to reduce negative outcomes in the midst of a crisis response.

Given its direct impact on life safety, gaining a greater understanding of the concept and practice of mitigation is indeed a worthy goal. Yet mitigation efforts are universally recognized as being underfunded and underutilized across the globe. To paraphrase a fellow emergency manager, "mitigation isn't sexy and it doesn't capture headlines. But it matters at every single incident scene."

It is also worth noting that mitigation is often a regional effort impacting rural, suburban, and urban communities. It is vital that governmental jurisdictions work across such divisions to minimize the impacts of truly large-scale disasters such as earthquakes and hurricanes.

As discussed in previous chapters, two key pieces of federal legislation in the United States that sought to attain such a worthy objective were the Stafford Act of 1988 and the Disaster Mitigation Act of 2000. Yet as with other phases of emergency management, little of the existing guidance for practitioners or legislative language focuses specifically on the urban environment. Much of the existing information regarding mitigation is highly strategic in nature.

Published in November of 2011, the US National Preparedness System (NPS) builds upon previous documents such and the National Response Plan and National Response Framework. However, as its title implies, this newer directive seems to be focused much more broadly than response alone. The NPS identifies five mission areas, defined as prevention, protection, mitigation, response, and recovery. These clearly align with the overall phases of emergency management. Crucially, the newer NPS guide also identifies defining risk as a key component, stating "a risk assessment collects information regarding the threats and hazards, including the projected consequences or impacts" (2011, p. 2). The NPS also introduces the Threat and Hazard Identification and Risk Assessment, or THIRA, which has quickly become a crux of urban emergency-management and homeland-security planning, and is directly tied to federal funding in the United States. Finally, the NPS presages the specific frameworks in each of the five mission areas—closely aligned with phases—of emergency management, including mitigation, which have since been released.

A number of authors and scholars have examined and assessed mitigation from a policy perspective. Sylves notes that once a risk to health or safety is identified, mitigation is essentially the decision to implement a "risk reduction program" (2008, p. 21). To differentiate two related phases, Sylves defines mitigation as "identifying and attempting to reduce hazard risks and vulnerabilities capable of producing disasters or emergencies" and preparedness as "anticipating and developing a variety of resources for response and recovery" (2008, p. 114–115).

As aforementioned, the US federal government's attention to mitigation policy has ebbed and flowed over the years. While the FEMA first found life during the administration of Jimmy Carter, it is important to note that mitigation was not a focus of the agency for many years. The Stafford Act of 1988 made important strides in recognizing the multiple phases and requirements of emergency management. Yet not until James Lee Witt was named as the head of FEMA under President Bill Clinton did the agency truly turn its attention to mitigating the impacts of natural or manmade disasters (Sylves, 2008, p. 68–69). As a widely respected and transformative leader, "central to Witt's vision was an increased emphasis placed on disaster mitigation" (2008, p. 68). Witt was the former head of emergency management for the state of Arkansas and widely respected in the emerging field. As noted previously, important US legislation leading to this resurgence of focus upon mitigation includes the Stafford Act of 1988, the Volkmer Amendment of 1993, and later the Disaster Mitigation Act of 2000.

Witt's concept of the mitigation phase overlapped other phases of emergency management. Emphasizing this vision, FEMA actively worked to move flood-prone structures off of floodplains across the Mississippi, Missouri, and Ohio River valleys following the catastrophic flooding during his very first year overseeing the federal agency. (Also see the case study in Chapter 13, Recovery Phase.) This effort thus overlapped with the recovery projects occurring immediately after the flood waters receded. Witt argued that spending funds expeditiously during the cleanup process would prevent repeated expenditures on the same structures or neighborhoods in future floods. Time has proven him to be very much correct in that assessment.

Birkland's analysis of disaster-related legislative actions notes that "there is reason to believe that policymakers are coming to understand the causal factors involved in the damage and disruption done by a range of disasters" (2006, p. 154). Yet he also adroitly cautions that much learning remains to be done and expresses particular concern at the cyclical nature of the political process in the United States and Europe. There is strong evidence that many current administrations across Europe and North America are not particularly inclined to provide financial or administrative support for any projects that don't play well on camera or social media.

Yet from a simple cost-benefit analysis viewpoint, mitigation seems to be an excellent investment, offering policymakers maximum return on investment. Roberts notes that mitigation activities "decrease the damage disasters cause to lives and property by anticipating hazards and making communities more resistant to their effects." He asserts that the US federal system "favors response, and to some extent, recovery, over prevention and especially mitigation" and further that "all levels of government have added more programs to aid response than to provide mitigation" (2009, p. 127–130). Birkland concurs, noting that FEMA's demotion to a subunit of the Department of Homeland Security "has rendered FEMA's mitigation strategies much less potent than they were during the Clinton-Witt era" (2006, p. 114).

Some evidence supports a potential shift back toward an all-hazards mitigation approach. An independent study by the Multihazard Mitigation Council of the National Institute of Building Sciences found that the United States ultimately saves four dollars for every dollar spent on mitigation projects overseen by FEMA. This marks a fiscal return far exceeding the metrics of a typical federal program.

## ANALYSIS

### Mitigation Basics

FEMA's definition of mitigation as action taken reduce the exposure to and impact of an attack or disaster is again relevant here—"action" being the key term. Mitigation is not a static or placid state of affairs but is rather a constant subcycle of analysis and application. It can quite accurately be viewed as a microcosm of the larger emergency-management cycle.

The process of mitigation depends first and foremost on the identification of hazards and threats which could lead to risk for a given area or population. Once again, in an urban setting, such risks can be expected to be significantly elevated due to the concentration of people and of critical infrastructure.

Effective hazard identification and risk assessment are crucial. No two urban areas face precisely the same hazards, making an honest assessment a challenging but worthwhile endeavor. Identifying natural hazards may include looking at historical factors to map floodplains or gather earthquake data. Assessing human-caused hazards may require examining potential technological risks or identifying potential terrorist targets. (See Chapter 5, Assessing Risk in Urban Areas, for more about risk assessment.)

Santella, Steinberg, and Parks cite the infrastructure characteristics identified by such hazard and risk assessments as crucial, noting it is in the public's best interest that policy and budgetary decisions "related to critical infrastructure must be made within an objective framework that takes into account all available information including the dependencies and interdependencies within and across infrastructures" (2009, p. 410).

A number of preincident mitigation tools can then be employed by public-safety officials. These include land-use and zoning restrictions, altering or updating construction codes, and working closely with financial institutions and insurance providers to maximize a given jurisdiction's status. Homes or businesses may be relocated to higher ground away from waterways following a flooding incident. These processes may not sound particularly exciting, but they can have truly outsized effects during response and recovery efforts.

Physical barriers are often a form of mitigation. This may be true in relation to both human-caused and natural hazards. For example, a reinforced security gate or series of concrete bollards may lessen the risk and impact of a vehicular borne improvised explosive device (VBIED) for a high-value facility or critical-infrastructure asset. Raising the height of an existing levee by 2 feet may delay or minimize overtopping of the structure during severe flooding. Both actions would reduce risk and minimize exposure to a specific threat, thus meeting the core definition of mitigation (Fig. 11.1).

Intelligence assessment should be considered a vital tool in mitigating technological and human-caused hazards. Working with the local fire department to understand what chemicals are used in a given factory, or with law-enforcement partners to determine any potential targets of terrorists or extremists is time well spent by an urban emergency manager. (In the United States, local fusion centers can be particularly helpful with the latter task.) Accurate information is at the core of a successful mitigation plan.

Ultimately, a best practice for mitigation is to follow the all-hazards model. If it is a risk to your organization or jurisdiction, it should be assessed, categorized, and mitigated against. Not all hazards or threats can be prevented, so mitigation will be vital to minimizing damage and compressing the chaos when major incidents do take place.

## Mitigation Tools

While the analysis of the prevention phase generally focused on human-caused hazards, the mitigation phase ultimately tilts toward natural hazards. It is noteworthy that many of the more successful techniques for natural hazards are also applicable to manmade or hybrid hazards. Once a holistic risk analysis has been accomplished, an organization or jurisdiction can then apply hazard-specific mitigation tools (Haddow et al., 2008, p. 77). Here, in alphabetical order, are several proven mitigation tools and methodologies available to urban emergency managers.

### *Data Modeling*

Gathering statistical data during hazard and risk assessments can inform later decision making. (See Chapter 5, Assessing Risk in Urban Areas, for more

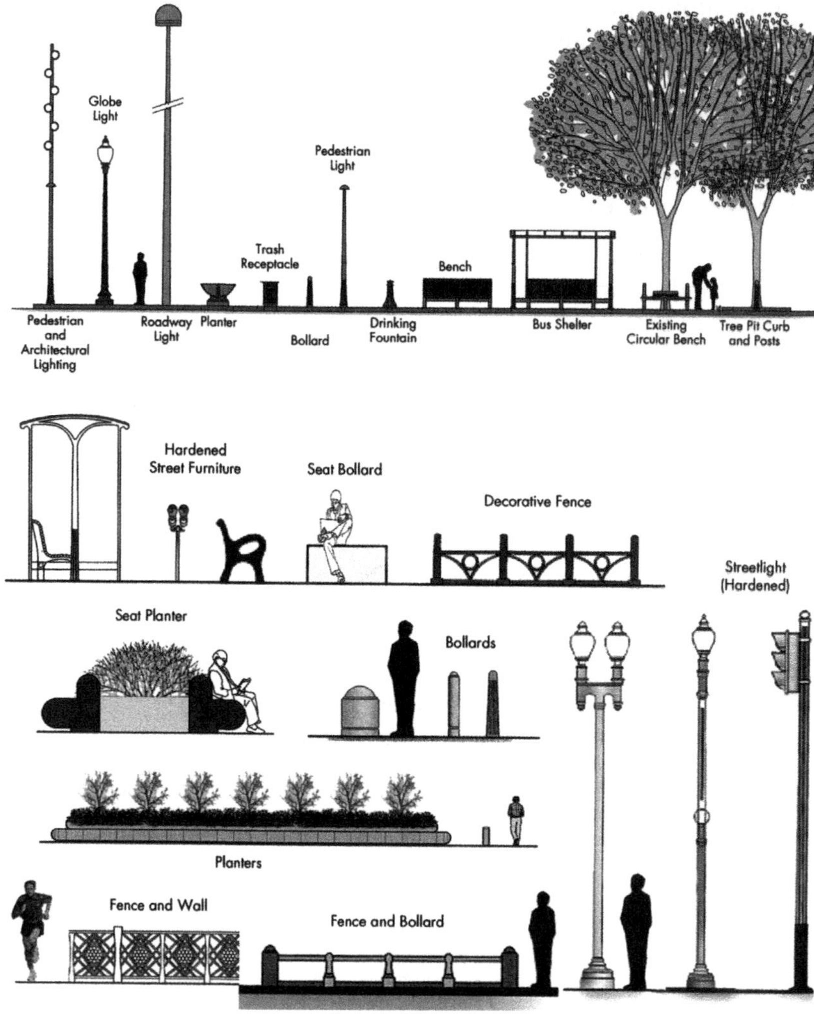


**FIGURE 11.1**
A wide variety of surface level vehicular barriers provide mitigation options for accidental or intentional vehicle-induced threats. *Source: FEMA.*

information.) Tracking infrastructure locations, functions, and interdependencies can help prioritize and guide mitigation efforts. Armed with criticality and geographical data, emergency managers are better able to apply mitigation policies and strategies across an organization or jurisdiction. Useful examples would include predictive floodplain mapping products or critical infrastructure interdependency graph models.

### *Design and Construction Codes*

As one of the most cost-effective methods of mitigating risk, instituting building and construction codes for new or renovated structures is worthy of serious consideration. Examples include retrofitting existing structures to stronger seismic standards or using fire-resistant materials for roofing and siding on residences near the wildland/urban interface. Regulation of vegetation or stormwater drainage serve as examples of fire and flooding mitigation, respectively.

### *Financial Incentives*

Many local governments are utilizing special tax assessments, bonds, relocation incentives, and assistance with federal grant applications as tools to reduce exposure to specific hazards. These tools may be used to encourage residents to relocate from low-lying, flood-prone areas, for example, or to financially incentivize seismic retrofitting of businesses storing hazardous chemicals. There is also a mitigation element to incentivizing nonstandard power sources such as solar arrays or back-up generators.

### *Incident Response Mitigation*

As noted in the beginning of the chapter, mitigation can occur across multiple phases of emergency management, including response. Ultimately, an incident commander or ranking officer has the opportunities in the field to make decisions that support mitigation. Examples would include firefighters utilizing water streams to douse structures adjacent to an active fire or officials using a mass-notification system to notify residents of a specific neighborhood of a utility outage. In each case, such mitigation efforts within the response phase may serve to limit or reduce the cascading damages and impacts of a given threat or hazard.

### *Insurance*

Traditional insurance policies transfer risk from residents or businesses to private-sector insurance providers. The government equivalent would be the National Flood Insurance Program (NFIP), which provides subsidized and affordable insurance in areas where traditional insurers may be reluctant to issue coverage. While it has been under significant price pressure in recent years, the NFIP provides the additional benefit of generating very detailed floodplain and riverine mapping products useful across multiple jurisdictions and sectors.

### *Land-Use Planning*

This mitigation tool relates to local development governance, including property sales, annexation, environmental review, and easements. In severe cases, local governments may use federal grant funding to purchase and demolish properties that have been repeatedly, catastrophically damaged by the same hazard. (See the case study in Chapter 13, Recovery Phase.) Related regulation may be used to protect specific animal or plant habitats and historic sites, or to control residential density.

#### *Policies and Procedures*

Perhaps the least expensive and most overlooked mitigation option is that of enforcing existing policies and procedures. Very frequently, mitigation measures such as floodgates or alternative power sources are not effectively tested or maintained. Establishing firm organizational or jurisdictional protocols, and making mitigation a topic of discussion at operations meetings or tabletop exercises, is an extremely cost-efficient method of maximizing measures already in place.

#### *Structural Controls*

Structural controls are primarily intended to reduce the impact of water flow, whether flood related or coastal erosion. Levees are by far the most common type of structure constructed under this mitigation option—with decidedly varied results. Breakwaters and seawalls may be utilized in coastal areas to reduce erosion along sections of coastline or to protect vital harbor infrastructure. Extensive environmental and hydrological studies are generally required prior to instituting or significantly modifying any such structures.

### National and Interjurisdictional Level

One of the most important elements of any detailed analysis will be just how to define urban mitigation in this context. Establishing such "ground rules" serves to better identify a manageable problem set.

One of the primary questions becomes: just what is mitigation in the realm of urban public safety? And here it is useful to further refine some overarching definitions. In the realm of mitigating natural or human-caused disasters, the FEMA definition seems highly relevant. The US federal agency identifies mitigation as "actions taken to reduce the exposure to and impact of" a given hazard (2003, p. B25).

As a specific example, one could cite the human-caused hazard of theft or vandalism. The deterrent effect of security cameras or fencing around a potential target mitigate the immediate threat, while gathering intelligence on a domestic terrorist organization potentially targeting a specific target likely falls outside of the recognized limits of mitigation in an emergency-management context.

Indeed, any complete threat or vulnerability analysis will look at any and all potentially damaging factors that confront a given facility or organization.

The three phases of prevention, preparedness, and mitigation would include making sure that the new fire extinguisher is positioned and charged correctly or replacing the faulty electrical outlet so that a fire is less likely to occur in the first place. (And any thorough fire-protection plan would doubtless include both.)

Yet there remains a significant focus on the phases of response and recovery at all levels of government. This is understandable for a number of reasons. Response and recovery are highly visible, and often politically charged, phases of the emergency-management process. Whether in Seattle, Sheffield, or Singapore, the local television station will carry footage of firefighters battling a multiple-alarm fire long before they will air scenes of a new storm drain being installed to control flooding.

But amounts of media attention do not necessarily translate directly to importance. It is safe to presume that the residents of New Orleans would have been quite interested in some improved mitigation measures—such as stronger levees and pumps—in the summer of 2005. (See the case study in Chapter 2, Development and History of Emergency Management.)

Thus urban emergency managers cannot fall into the "either/or" trap—they must seek a balance of "both/and" to be successful. No single emergency-management strategy will succeed by standing alone, just as no phase exists in a vacuum.

Present mitigation-specific programs at the federal level are quite sparse in American and European cities. FEMA administrator Craig Fugate, having been appointed by incoming President Barack Obama in 2009, placed a renewed emphasis on mitigation within the federal emergency-management establishment. His experience as a state leader in hurricane-prone Florida unquestionably taught him valuable lessons.

Indeed, separate grant programs for predisaster mitigation and postdisaster mitigation do exist in the United States, generally as extensions of the Stafford Act. However, these programs are heavily focused on natural disasters, and flooding in particular. This prioritization by FEMA is hard to criticize given the comparative frequency, and cumulative damage costs, of flooding in comparison to acts of violence or terrorism.

Two of the larger federal efforts are the Hazard Mitigation Grant Program (HMGP), which is focused on flooding, and the Predisaster Mitigation Program, which as the name implies is designed to place mitigation efforts prior to response in the ongoing emergency-management cycle. Similar programs focus on other specific hazards such as earthquakes and hurricanes. This sampling does accurately portray the struggle between the mitigation of natural versus human-caused hazards that continues within the US federal establishment.

Two additional programs that are worthy of further examination are the Urban Area Security Initiative (UASI) and the Flood Mitigation Assistance (FMA) program. Combined, these programs provide excellent examples of mitigation for human-caused and natural hazards.

UASI is a large and diverse grant program that—as implied by its name—focuses on large urban areas. At its peak, 64 of the largest metropolitan areas in the United States received some level of funding under the program (although there is ongoing legislative pressure, led by New York City, to reduce that number to 10 or so). The funding can be utilized for all phases of emergency management and is ideally shared among multiple jurisdictions and departments in a given geographical region. As a point of reference, $834 million was allocated by the federal government in FY2010, according to FEMA's Responder Knowledge Base (RKB). When properly justified, multimillion dollar allocations are not uncommon for large-scale projects such as interoperable communications networks or interlinked digital video cameras.

As natural hazards continue to evolve, flooding remains the greatest financial threat across the Western world. In the United States, the FMA program provides funds for projects to reduce or eliminate risk of flood damage to buildings that are insured under the NFIP. In practice, such efforts may involve supporting wetlands to help soak up flood waters, building or enforcing levee and drainage systems in populated areas, or buying out properties that have been repeatedly damaged by floods.

Similarly, Europe has faced extensive flooding damage from 2010 to 2014 and is considering a broad package of local and regional mitigation packages. The LIFE Regulation Program instituted by the European Union is another example of a large-scale hazard-specific effort—in this case seeking to mitigate the impacts of climate change and sea-level rise.

Regardless of any success or traction these programs have achieved, they are under dire funding threat as a result of the overall economic and political woes. UASI has seen enormous cuts to its funding, with a roughly 60% reduction in overall federal allocations between FY2009 and FY2012. This downward pressure has continued into the second half of the decade. The FMA program is consistently and chronically underfunded, receiving emergency infusions of cash following events such as 2012's Hurricane Sandy on the eastern seaboard of the United States.

Yet federal entities such as FEMA and the Department of Homeland Security (DHS) can also lead by identifying best practices and sharing information among state and local agencies—not just by providing grant-specific funding. Perhaps the most concrete example of this established strategy is the National Infrastructure Protection Plan, or NIPP.

The NIPP was first released in 2006, with a significant revision in 2009. It focuses on a "big picture" view of critical infrastructure and key resources (CIKR) across the United States. The emphasis within the NIPP is clearly regional systems, failures of which would directly impact large numbers of US citizens or overall national security. At 188 pages, its heft can be off-putting to some potential users; however, many segments are actually quite nuanced and flexible.

And yet, an anecdotal gap analysis among state and local jurisdictions shows that this document is not widely utilized by emergency managers or planners. This observation touches upon a common theme: that potentially useful tools have been created at the federal level, only to go unnoticed or unimplemented by state and local jurisdictions. Much more widely utilized is the FEMA Emergency Management Institute (EMI) and its Independent Study Program. This online series of courses includes brief but useful introductions to concepts such as CIKR and the NIPP, among others.

Finally, for several years, the DHS Office of Infrastructure Protection has been executing an annual "datacall" information-gathering process. This process has utilized state and local stakeholders to gather data about CIKR and submit it to a federal database via an online portal. Similar to the NIPP, the DHS datacall is significantly focused upon large-scale systems with significant regional or national implications.

Thus, the status quo—at least at a US federal level—relies heavily on providing grant funding for local and state mitigation projects. While efforts are underway to both "push" best-practices documentation out to this audience and to "pull" CIKR data from them, the overall rate of long-term success has yet to be determined.

## Local and Private-Sector Level

While national programs can set the overall tone, programs at the corporate, local, and state level can be just as crucial to mitigation. In the United States and Europe, grant funding tends to follow a "top down" model—emanating from federal or national governments or from the European Union.

However, a large percentage of the actual project management occurs utilizing a "bottom up" model such as the UASI program cited previously. In that case, projects are directly overseen by local or state emergency managers who are ultimately responsible for implementing physical-security mitigation efforts.

From an infrastructure standpoint, it is highly likely that local or private-sector officials will have greater insight into existing risks and needs. Again, the crucial first step will be to justify such concerns or perceptions with a thorough risk and threat assessment program. Only then can vertically integrated mitigation projects be launched and sustained.

Companies and nonprofit entities are partnering with local jurisdictions to plan for future emergencies. The city of Philadelphia has recently partnered with local electrical utility PECO to install new monitoring equipment to help map usage and outages and has begun identifying specific emergency-generator requirements for key structures and facilities (McIntire, 2015). Much of this collaboration was motivated by the impacts of recent storms and by data and research showing the jurisdiction to be at a high risk of future extended outages. (Additional examples may be found in Chapter 7, Collaboration in Urban Environments.) Being able to provide data and metrics to senior management or elected officials is a crucial component to validating the importance of mitigation.

Emergency managers in urban environments must take the lead in developing, driving, and sustaining mitigation efforts within their organization or jurisdiction. They must use every tool at their disposal in order to convey the value and criticality of minimizing negative impacts during times of true crisis.

## CASE STUDY: CHICAGO'S INFRASTRUCTURE DATABASE

As an example of physical-security and infrastructure mitigation, Chicago is quite similar to other large, urban jurisdictions in the United States. The city is the third largest in the nation with a population of approximately 2.7 million people and has a dedicated emergency-management department, the Office of Emergency Management and Communications (OEMC). In coordination with the Chicago Police Department and Chicago Fire Department, emergency managers contribute to regular updates of an all-hazard mitigation plan.

Locally funded hazard-mitigation projects are generally scheduled only as a portion of "regular" updates or maintenance, which would include public-health measures or physical-security upgrades, or due to specific intelligence threat information. Due to restricted budgets on both the local and state level, very little funding is available from either tier of government. As a result, virtually all substantial projects are funded primarily via federal grants.

On the ground in Chicago, a group of subject-matter experts collaborates on virtually every project of this type. This group vitally includes federal infrastructure experts known as Physical Security Advisors, or PSAs. Specialized local law-enforcement, fire services, and emergency-management personnel participate on an as-needed basis. Individual locations or subsystems can request site-assistance visits, which include a basic threat and vulnerability assessment.

The actual completion of any physical-security mitigation project under these programs is the sole responsibility of the jurisdiction. In Chicago, this task falls to the OEMC and its designated contractors. Typical applications include blast-resistant windows and trash receptacles, vehicular barriers, networked video cameras, and specialized equipment for responders. The vast majority of these expenditures are clearly mitigation-related. While the program is still inconsistent and less than holistic, it has the potential to address a wide range of infrastructure and assets.

In addition to the aforementioned hazard-mitigation document and various strategic and policy planning, the City of Chicago maintains the largest and most detailed critical-infrastructure database in the United States. As of 2014, the Facility Information Management System (FIMS) indexes over 8000 locations.

FIMS warehouses information crucial to responders and planners such as floor plans, contact numbers of key personnel, and crucial hazardous-materials data. The database itself is largely populated by private-sector stakeholders who own or operate vital systems, which is then vetted by the local fusion center. (See Chapter 8, Emergency Management and Homeland Security in Urban Areas, for additional detail.) This information can then be rapidly accessed by command staff arriving on an incident scene.

From a mitigation perspective, FIMS provides a unique platform to search by address or latitude/longitude coordinates to identify all critical locations within a given geographical area. The Chicago/Cook County UASI and the State of Illinois are also considering databases interconnected and/or built upon the FIMS platform, potentially making the system a regional resource for such planning activities.

Funding and bureaucracy remain challenges, as always. While federal grant revenues continue to fall, and with local and state budgets for such mitigation measures virtually nonexistent, financing becomes a significant hurdle. The average mitigation project requires dozens of pages of planning documentation, much of it redundant information presented in different formats for different agencies.

It will thus be incumbent on Chicago (and other cities) to spend the money it has very wisely, as—at the very least—a significant interruption in federal funding may be anticipated in the months and years ahead. In terms of physical-security mitigation at the local level, "doing more with less" is not a cliché, it is an absolute necessity.

## CONCLUSIONS AND IMPLICATIONS

From this discussion and analysis, certain discernable patterns begin to emerge. Ultimately, several best practices for all-hazards mitigation efforts within the urban landscape can be identified.

### Consider All Potential Hazards

It is a fact that a lot can go wrong in any given densely inhabited jurisdiction. An honest assessment of potential hazards, threats, and vulnerabilities is an absolute necessity. An all-hazards threat analysis is a time-consuming and work-intensive undertaking. It is also the bedrock of emergency-management planning for any city, and the basis for all that follows. Information is power, particularly when it comes to mitigation.

### Know What You've Got

Identifying all of the potential hazards is a necessary first step. Knowing what is worth protecting is also crucial. A thorough listing of all CIKR within a jurisdiction is a logical next step. The FIMS database in Chicago can serve as a model for this process, but for many organizations or jurisdictions, thousands of entries would simply be overkill. Begin by identifying the absolutely critical systems and facilities that a given city cannot function without, and you've effectively identified the first locations worthy of targeted mitigation efforts.

### Connect All Levels of Governance

To be truly successful, any all-hazards mitigation efforts must be both horizontal and vertical. This means developing a good relationship with stakeholders at the local, state, federal, and possibly regional or tribal levels. In addition, use subject matter experts from inside and outside the jurisdiction. Maintain good relationships among specialists from groups such as the local SWAT team and hazardous-materials response unit. These sorts of relationships are invaluable as a "champion" prepares to explain why a given project is worthy of planning efforts and funding dollars.

### Be Flexible With Funding Sources

As has been discussed, the foreseeable future will be a difficult time to secure funding for substantial mitigation projects. Federal grants are being slashed, and many state and local jurisdictions are in a period of flat or declining budgets. Creativity is thus at a premium, and local planners should be unafraid to pursue tough price negotiations with potential contractors or to seek funding from outside sources. For example, one local jurisdiction developed a security-mitigation plan for a key financial institution in the city, and the project itself was then funded wholly by private-sector financing. An effective plan should be highly justifiable from a cost-benefit perspective. In essence, if a site or system is worth protecting, a method can be found to provide the funding to do so.

### Imitation Is the Highest Form of Flattery

No single jurisdiction has all the answers. One of the most refreshing characteristics of the young field of disaster science is its willingness to adapt, borrow, and/or steal good ideas from elsewhere. Excellent, cost-effective, successful programs are being instituted in cities across the developed world, and it is incumbent on a good emergency-management planner to keep abreast of any such developments. Other fields of study can offer key insights as well, from the behavioral analysis of sociology to the latest developments in the physics of blast-mitigation materials. There is no need to reinvent the wheel—particularly if someone further along the road is willing to share their design.

## SUMMARY

This chapter has introduced several key terms, provided a brief case study, examined the current state of all-hazard mitigations efforts in the United States in particular, and suggested several potential best practices going forward. Whether the threat comes from a lone-wolf terrorist organization or from a major storm system, reducing the opportunity or impact of loss to persons and property is indeed a worthy pursuit.

The NPS released by DHS indicates that a strategic change may be occurring in the way that the federal government perceives and guides emergency-management and mitigation practices within the United States. It is clear that all-hazard mitigation is worthy of further attention, assessment, and refinement by emergency-management students and practitioners.

## References

Birkland, T. A. (2006). *Lessons of disaster: Policy change after catastrophic events.* Washington, D.C: Georgetown University Press.

Haddow, G., Bullock, J., & Coppola, D. (2008). *Introduction to emergency management.* Burlington, MA: Butterworth-Heinemann.

McIntire, M. E. (2015). How the most at-risk cities are preparing for more power outages. *Emergency Management.* Retrieved from: <http://www.emergencymgmt.com>.

Roberts, P. S. (2009). A capacity for mitigation as the next frontier in homeland security. *Political Science Quarterly, 124*(1), 127–142.

Santella, N., Steinberg, L. J., & Parks, K. (2009). Decision making for extreme events: Modeling critical infrastructure interdependencies to aid mitigation and response planning. *Review of Policy Research, 26*(4), 409–422.

Sylves, R. (2008). *Disaster policy & politics: Emergency management and homeland security.* Washington, D.C: CQ Press.

## Further Reading

Department of Homeland Security. (2009). *National Infrastructure Protection Plan.* Retrieved from: <http://www.dhs.gov/xlibrary/assets/NIPP_Plan.pdf>.

Department of Homeland Security (2011). *National Preparedness System.* Washington, D.C: U.S. Federal Government.

Federal Emergency Management Agency (2003). *Reference manual: To mitigate potential terrorist attacks against buildings.* Washington, D.C: Federal Emergency Management Agency.

Federal Emergency Management Agency (2005). *Risk assessment: A how-to guide to mitigate potential terrorist attacks against buildings.* Washington, D.C: U.S. Federal Government.

Federal Emergency Management Agency. (2011). *Responder Knowledge Base.* Retrieved from: <www.rkb.us>.

Multihazard Mitigation Council. (2005). Natural Hazard Mitigation Saves. Retrieved from: <www.nibs.com/resources>.

Webster's Dictionary. Mitigate. Retrieved from: <http://www.webster-dictionary.org/>.

CHAPTER 12

# Response Phase

## INTRODUCTION

November 30, 2007 was a cool but clear day in Chicago. Just before noon, Amtrak passenger train 371, consisting of one diesel locomotive and three passenger cars, was approaching its final stop at Chicago's Union Station. Many of the nearly 200 passengers were putting away personal items and gathering their bags as the train, originating in Michigan, was merely a few dozen blocks from its final destination.

As the train moved through a railyard just miles south of Union Station, the engineer apparently misread a signal at the Englewood Interlocking junction. The color-coded signal indicated a track speed limit of 15 miles per hour and warned crews to be prepared to stop. A federal report would later state that the engineer misinterpreted the signal and was in fact traveling through the railyard at 30 miles per hour. A Norfolk Southern freight train was completely stopped on the same track just ahead.

Despite applying full emergency braking, the Amtrak train struck the motionless freight train with enormous force, causing the diesel locomotive to leap airborne and literally grind to a halt on top of the last boxcar in the freight train. Quite remarkably, the locomotive remained upright and did not burst into flames. Unsurprisingly, the massive impact caused extensive injuries to the passengers and crew of the Amtrak passenger train.

At approximately 11:24 a.m., 911 calls began to come in to the Chicago Office of Emergency Management and Communications (OEMC), which functions as the jurisdiction's emergency call taking center and the home of its emergency managers. A large number of police, fire, and emergency-management assets made their way to the crash site with lights flashing and sirens blaring. They were in route to a surreal scene of one train sitting partially atop another with more than 70 seriously injured persons—a true mass-casualty incident.



Urban Emergency Management. DOI: http://dx.doi.org/10.1016/B978-0-12-805307-2.00012-3

Critically, the railyard operator had shared its security-camera feeds with the city's Operations Center, also located at OEMC. As a result, dispatchers there had a real-time view of the crippled train and the surrounding terrain. They were able to ascertain which direction of approach would be quickest and safest for responders and to guide them in via radio and text messaging.

The accident site was atop a roughly 20-foot raised berm in a residential neighborhood on Chicago's near south side. Multiple rail lines ran parallel to the west, making approach from that direction extremely dangerous for responders. A security fence on the east side of the accident site obstructed responders momentarily, with the first units climbing up a Chicago Fire Department (CFD) aerial ladder deployed from street level to clear the fence. This obstruction was soon circumvented with the arrival of several CFD saws-all apparatus.

As police units established a perimeter, fire-services, paramedic, and emergency-management units performed a quick walk-around assessment of the scene. Multiple passengers and crew members were injured and in need of triage. Diesel fuel was spewing from the crippled Amtrak locomotive, perched precariously on top of the boxcar upon which it landed. A rapid decision was made to stabilize and evacuate all civilians from the crash scene as quickly as possible because of ongoing and immediate threats.

At the foot of the berm, downhill at street level, a massive triage operation sprung to life. Firefighters made entry into the passenger cars, climbing over broken seats and scattered luggage. Emergency-management personnel led the walking wounded to a nearby school gymnasium, where they could be temporarily sheltered from the freezing temperatures. Paramedics began assessing the seriously injured as they emerged from the train, lining up dozens of ambulances in a rapid "load and go" operation.

Multiple passengers with minor injuries had been gathered from the school and transported via buses to select local hospitals which had not received the more grievously wounded. Uninjured passengers were bused to Union Station to complete their journey and reunite with loved ones. Within 3 hours, the entire scene was secured and turned over to federal investigators from the National Transportation Safety Board (NTSB) and the Transportation Security Administration (TSA).

Luck—good or bad—also plays an undeniable role in any such large-scale response. The Amtrak locomotive landed upright atop the last boxcar of the parked freight train, rather than rolling to either side. A separate freight train on an immediately adjacent track had multiple tanker cars containing a range of liquid cargos, including hazardous materials; these did not ignite or develop any significant breaches. Hundreds of gallons of diesel fuel leaked from the crippled Amtrak locomotive; due to diesel's relatively high ignition temperature, this spill also failed to ignite. Amazingly, despite 71 injured passengers and crew, no lives were lost.

Such urban responses occur on a regular basis. As a new emergency manager with less than a year on the job when responding to this scene, the author has yet to see a logistical response of the magnitude witnessed that day. The dispatchers who guided units in, the firefighters who climbed aboard the crippled train, the officers who held the perimeter, and the paramedics who ran a flawless mass-casualty triage and transport operation, all embody the true essence of emergency response.

## ANALYSIS

As the most visible phase of the emergency-management cycle, response receives an outsized proportion of attention from the public, the media, and elected officials. Response is the piece that includes dozens of cameras and satellite trucks, and that generates Congressional hearings.

This phase of emergency management may generally be defined as the period during which emergency personnel are deployed into the field in reaction to a developing incident. Response can be as a result of natural, human-caused, or hybrid emergencies.

But what is an effective response? How does an organization or jurisdiction succeed in a crisis situation?

### Plan First, or Plan to Fail

The key first step in effective response is, not surprisingly, effective risk assessment and emergency planning. (Additional guidance may be found in Chapter 5, Assessing Risk in Urban Areas.) A holistic emergency operations plan (EOP) will provide responding agencies with a common playbook when units are deployed to the field.

Erickson (2006, p. 67–68) cites five core principles of a thorough response plan:

- Facility owners and operators should identify and plan for known risks
- Training must accompany and reinforce any plan
- Communications will be key to any response
- Audits and drills should lead to regular plan refinement and updates
- An established, on-site chain of command is vital

The portion of an EOP related to response is truly the core of the document, addressing the portion of the emergency-management cycle in which lives and property are in direct and immediate danger. Crucial components will include defining specific responder tasks, and identifying key skillsets and communication requirements for each task (Erickson, 2006, p. 87).

## All Disasters Are Local

One of the core tenants of the incident command system (ICS) is that all emergencies or disasters be handled by the lowest-level agency with the capability and capacity to do so. This translates to the vast majority of emergency situations that occur on a daily basis being handled by the organization or company where they take place, often with the assistance of local first responders.

As an example, imagine a chemical spill at a local manufacturing plant. The first tier of response would come from workers and supervisors within the facility. As outlined in their emergency plans, they would work to contain and mitigate the spill and to determine any ill effects it may have on the health of employees or on the facility itself. If the threat was too large to be fully contained by plant personnel, they would then seek the assistance of local hazardous materials responders—whether through the local fire department or via contracted specialists.

Such occurrences take place on a daily basis in urban environments across the globe, and rarely involve state, regional, or federal authorities until the reporting portion of the recovery process. This is the key concept of an emergency being addressed by the most localized resources capable of doing so. Thus, the most highly skilled and highly equipped units may be dedicated to providing guidance to less specialized units and deployed into the field only when truly required by an urgent or multijurisdictional threat.

It is also worth noting that emergency-management assets are frequently deployed during large-scale preplanned events such as parades or professional sports matches. In the parlance of emergency management, an event is an organized and preplanned occurrence, while an incident is an unplanned or unexpected emergency.

## Effective Response

The response phase can be especially chaotic and fraught with misinformation. True response is many functions rolled into one—providing critical emergency assistance, reducing the possibility of secondary or cascading damage, and identifying challenges and opportunities for recovery operations (Sylves, 2008, p. 23) (Fig. 12.1).

As with all phases of disaster, the core priorities remain protecting human life, infrastructure, and the environment. Responder safety must also be

**FIGURE 12.1**
Chicago Fire Department (CFD) personnel battle a three-alarm commercial fire in November of 2008.
*Credit: Author.*

viewed as an overarching responsibility of incident command. The response plan should also be evaluated and adapted as conditions change in the field (Bevelacqua & Stilp, 2009, p. 7). Ultimately, all emergency managers seek to compress and minimize the chaos that results from a major emergency or disaster. An effective use of resources, prompt transition to the recovery phase, and a resumption of relative normalcy may be viewed as longer-term goals.

## Whole Community Response

Each community or organization must identify individuals or facilities who may require additional or accelerated assistance during the response phase. During such situations, populations with a wide range of special needs may require significantly higher levels of manpower per capita, and responding agencies must be kept aware of such individuals. As noted in Chapter 10, Preparedness Phase, a partial list would include:

- Children
- The elderly
- The chronically ill or hospitalized
- The disabled
- Immigrant populations
- Non-English-speaking populations
- Minority populations
- The poor or homeless
- The illiterate
- The mentally ill

- Prisoners
- Transitory populations such as college students or hotel guests

## KEY COMPONENTS

Emergency response is, by definition, rapid and haphazard in its initial phases. Even with effective notification and dispatch systems, not all law enforcement, fire services, or emergency medical units will arrive at the same time or in perfect order.

Yet over many years of responses varying in timing, severity, size, and location, a number of best practices have emerged for effective utilization of resources. By studying past incident responses, a number of key components for emergency managers may be identified.

### Urban Response Types

First, a disclaimer: there is no average day at the office for an urban emergency manager. You may have had a neighborhood outreach session scheduled when a train derails on the opposite side of your jurisdiction. You may have had a mitigation grant funding meeting planned when a large warehouse fire erupts a few blocks away.

Flexibility is a key personality trait of the most effective emergency and crisis managers. In a complex urban environment, there are so very many things that can go wrong. Percentage wise, presuming that there will be an emergency on any given day does not make you a pessimist—it makes you a realist.

#### *Natural Hazards*

A wide range of natural hazards are arrayed against most urban settings. These certainly vary geographically and by season, but several are common enough to warrant review. As discussed in Chapter 3, Natural Hazards and the Urban Environment, such hazards may include:

- Earthquakes
- Flooding
- Hurricanes
- Pandemics
- Storm Surges
- Tornadoes
- Tsunamis
- Wildfires

Weather is the overwhelming commonality here. Too much or too little water is a recurring threat to major urban areas across the planet. Severe storms are common across much of the world, and large cities are traditionally located along waterways which may be prone to flooding.

Finally, there are a handful of outlying "black swan" events which are very hard to predict. These would include earthquakes which sometimes result in tsunamis, and the widespread and rapid spread of disease referred to as a pandemic.

### *Human-Caused Hazards*

People are often quite good at causing havoc themselves, as was reviewed in Chapter 4, Human-Caused Hazards and Hybrid Hazards. Manmade crises include but are not limited to:

- Active Threats
- Cybercrime
- Hazardous Materials Releases
- Infrastructure Failures
- Structure Fires
- Terrorism

Such disasters have human beings as the key and core causative factor. Cybercrime and terrorism are somewhat unique in that they represent a cognitive and capable opponent rather than a nascent hazard. Other types of human-caused hazards are generally a result of a failure of design or action—think of a bridge collapse or leaking tanker car (Fig. 12.2).

### *Hybrid Hazards*

Hybrid hazards are more difficult to define since they, by nature, have more than one cause. Examples could include a weather event that damages or degrades critical infrastructure, or an earthquake that causes a chemical release. The Fukushima catastrophe referenced in Chapter 4, Human-Caused Hazards and Hybrid Hazards, is the largest scale hybrid disaster in recent years.

Also worthy of mention is the gathering scientific evidence that climate change is likely to contribute to multiple future hybrid disasters, primarily due to higher sea levels. Such changes will magnify the impact of coastal flooding, tsunamis, storm surges, and hurricanes.

## Responding Agencies and Emergency Management

A range of people, organizations, and agencies will respond to a major emergency. As mentioned, all disasters start as very localized events. The true first

UNCLASSIFIED

**Bomb Threat Stand-Off Distances**

| Threat Description | Explosives Capacity[1] (TNT Equivalent) | Building Evacuation Distance[2] | Outdoor Evacuation Distance[3] |
|---|---|---|---|
| Pipe Bomb | 5 LBS/ 2.3 KG | 70 FT/ 21 M | 850 FT/ 259 M |
| Briefcase/ Suitcase Bomb | 50 LBS/ 23 KG | 150 FT/ 46 M | 1,850 FT/ 564 M |
| Compact Sedan | 500 LBS/ 227 KG | 320 FT/ 98 M | 1,500 FT/ 457 M |
| Sedan | 1,000 LBS/ 454 KG | 400 FT/ 122 M | 1,750 FT/ 533 M |
| Passenger/ Cargo Van | 4,000 LBS/ 1,814 KG | 600 FT/ 183 M | 2,750 FT/ 838 M |
| Small Moving Van/ Delivery Truck | 10,000 LBS/ 4,536 KG | 860 FT/ 262 M | 3,750 FT/ 1,143 M |
| Moving Van/ Water Truck | 30,000 LBS/ 13,608 KG | 1,240 FT/ 378 M | 6,500 FT/ 1,981 M |
| Semi-Trailer | 60,000 LBS/ 27,216 KG | 1,500 FT/ 457 M | 7,000 FT/ 2,134 M |

This table is for general emergency planning only. A given building's vulnerability to explosions depends on its construction and composition. The data in these tables may not accurately reflect these variables. Some risk will remain for any persons closer than the Outdoor Evacuation Distance.

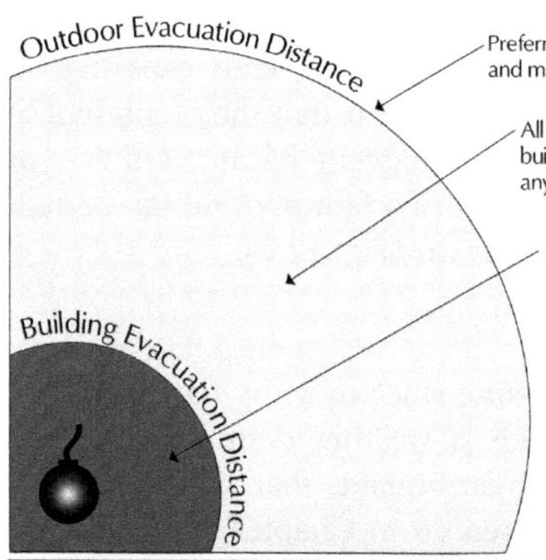


Preferred area (beyond this line) for evacuation of people in buildings and mandatory for people outdoors.

All personnel in this area should seek shelter immediately inside a building **away from windows and exterior walls**. Avoid having anyone outside—including those evacuating—in this area.[4]

All personnel must evacuate (both inside of buildings and out).

1: Based on maximum volume or weight of explosive (TNT equivalent) that could reasonably fit in a suitcase or vehicle.

2: Governed by the ability of typical US commercial construction to resist severe damage or collapse following a blast. Performances can vary significantly, however, and buildings should be analyzed by qualified parties when possible.

3: Governed by the greater of fragment throw distance or glass breakage/falling glass hazard distance. Note that pipe and briefcase bombs assume cased charges that throw fragments farther than vehicle bombs.

4: A known terrorist tactic is to attract bystanders to windows, doorways, and the outside with gunfire, small bombs, or other methods and then detonate a larger, more destructive device, significantly increasing human casualties.

NCTC

UNCLASSIFIED

000841ID 10-05

**FIGURE 12.2**

A widely utilized chart displaying the approximate indoor and outdoor shelter distances for explosive blasts of various sizes. *Source ATF/NCTC.*

responders will be civilians at the initial incident scene—employees, passers-by, visitors, contractors, etc. These individuals are not sworn to protect their fellow citizens, and yet many of them will put themselves in harm's way to do exactly that.

And many emergencies will be contained at this level—the trash can fire will be extinguished, or an injured employee will be transported by coworkers to the local hospital. If the capacity of a given facility or organization is exceeded, additional resources will be summoned to the location. This may come via automatic alarm systems, or via calls to local 911, 112, 999, or 000 emergency services requesting assistance.

### Local

In most major cities, calls for emergency services are routed to a call-taking and dispatch center. Often referred to as a public-safety answering point (PSAP), these facilities are a multidisciplinary point of first contact for citizens reporting an emergency in progress. Systems vary from jurisdiction to jurisdiction, but an emerging best practice is to collocate fire services, emergency medical, and law enforcement dispatchers in a single unified command facility.

The PSAP will gather as much information as is possible about the emergency, and then dispatch appropriate units to respond. It is becoming ever more commonplace for PSAPs to utilize additional sources of information, including live camera feeds, digital mapping products, or third-party alarm companies, to verify the veracity and scope of a given incident.

Local responders handle hundreds of thousands of calls for service on a daily basis. Many are simple and may be addressed by a single unit. Imagine a local fire engine crew performing a well-being check on an elderly neighbor, or a county sheriff taking a shoplifter into custody. This represents the average "hum" of an urban area addressing day-to-day concerns. Local responders utilize their own training, protocols, and procedures to dictate such deployments. Frankly, these professional units are very capable and competent at their jobs overall, and need little or no input from emergency managers. But there are exceptions.

Occasionally, an event or incident will be of such a scope and scale that police, paramedics, and firefighters need another level of responder. This may mean organizing an EOC for a large-scale outdoor gathering. In many jurisdictions, it may mean deploying directly to the field alongside other first responders.

In 2011, the author took an electronic poll of his fellow emergency managers from 30 large American cities, ranging in size from Cleveland, Ohio, with roughly 390,000 residents to New York City, with roughly 8.4 million residents. All had at least some full-time staff dedicated to emergency management within their jurisdiction. The core question of the poll was whether those emergency managers responded to incidents in the field. The results were split almost perfectly into thirds—one-third indicated no response, one-third indicated response only to major events or when dictated by senior elected officials, and one-third indicated regular field response.

This simple query is very telling for urban emergency managers. Different jurisdictions utilize this very valuable resource in different ways. Some cities view the emergency operations center (EOC) as the right location for their emergency managers, while others prefer them to be forward deployed to frontline command posts. Some large jurisdictions such as New York City, which has the largest full-time cadre of emergency managers in the nation, have the available personnel to perform both functions (and many additional tasks).

There is no absolutely correct answer, and today's emergency manager must continue to place himself or herself in the most useful and advantageous position during any given event or incident. Regardless of where the emergency manager is deployed, he or she should seek to make the greatest possible contribution to the response effort.

This process often involves functioning as a Liaison Officer, Planning Section Chief, or Logistics Section Chief within the incident command structure. (See Chapter 6, Incident Command Systems and NIMS, for more information.) It is crucial for a high-performing emergency manager to develop key relationships prior to a major event or incident occurring. These partnerships should be both vertical and horizontal. In other words, an effective public-sector manager should be very familiar with the two rungs above and below his/her position on the chain of command as well as fellow public safety officials from neighboring jurisdictions. An emergency manager in the private or nonprofit sectors should have similar connectivity among his or her peers.

### State/Regional

Very rarely, an event or incident will exceed the capabilities of local responders, or of their preexisting mutual-aid agreements. In such cases, state or regional assets or personnel may be called upon to assist. Various nations may refer to such geographic governing units as provinces, states, or regions.

These units are often overseen by a different layer or stratus of governance, and their assistance will only be provided via the specific request of a recognized local official. This is both to avoid overuse or reliance on such assets and also to coordinate larger-scale or regional disaster responses.

The classic example in the United States would be a case of flooding. In the spring of 2016, extremely heavy rains deluged some parts of the Houston metropolitan area with more than 15 inches of precipitation in a matter of hours. Local responders, including the very capable Harris County emergency-management agency, quickly recognized that their resources would be rapidly outstripped by such a large-scale weather incident. Texas state troopers were activated to provide traffic control around flooded roadways. The National Guard provided heavy equipment and personnel. Specialized state-level search and rescue teams and swift-water boat units were activated and deployed.

Such a rapid escalation of the threat requires an equally rapid escalation of the levels of governmental dedicated resources to the response phase. State and regional assets do not come cheaply, and there is certainly an expense that will result. However, this is more than offset by the reduction of risk to human life and property that these specialized units embody.

Most US states have their own emergency-management agencies, as do many regional governments in Europe and Asia. Emergency managers assigned to these departments generally have a wider geographic area of responsibility than do their local or corporate counterparts, yet their overall duties remain much the same. Western nations often use federal funds or grants to staff, train, and equip state or regional teams—largely in the goal of resolving as many emergencies at this level as possible rather than escalating to a national response.

If an activation occurs at this level, emergency managers should expect to be assigned as part of an area command structure. (See Chapter 6, Incident Command Systems and NIMS, for more information.) Multiple ongoing incidents and areas or operation should be anticipated, as should the involvement of multiple local jurisdictions. Such responses tend to be geographically broad, and extensive in terms of length of deployment.

### Federal/National

Exceedingly rarely, an event or incident in an urban area will be so extensive that local, state, and regional resources are simply unable to handle the tasks required. In such cases federal or national-level assets and personnel will be activated.

This actually occurs somewhat commonly on the law-enforcement side, when a high-ranking elected official or foreign dignitary visits a given jurisdiction. A protective detail of Secret Service agents, the Diplomatic Security Service, or US Marshals, for example, may work with state and local authorities to coordinate the safety of their protectee.

Yet such events rarely involve extensive efforts on the part of corporate, local, or state emergency managers. In this context, a federal-level event may be presumed to be an exceedingly large emergency or disaster affecting multiple jurisdictions. Modern examples would include the September 11 terrorist attacks, Hurricane Katrina making landfall in 2005, and the widespread European flooding in 2013.

While such disasters have very different causations, they have a common theme—local authorities can in no way mitigate or respond to such vast disruption. Such instances require prompt escalation via the varying layers of government to secure the most competent and qualified responders across a wide range of expertise.

In the United States, this means the Federal Emergency Management Agency (FEMA), which has significant authority—and a significant budget—at its disposal during declared emergencies. Here again, we find a layer of professional emergency managers who build upon the roles and skillsets of their corporate, local, and state counterparts. This layer represents the full extension of the "handle at the lowest possible level" model. Local officials formally sought state assistance, and state officials formally sought federal assistance. There are 15 core response and coordination activities recognized by the US federal government, referred to as emergency support functions (ESFs):

- ESF #1, Transportation
- ESF #2, Communications
- ESF #3, Public Works and Engineering
- ESF #4, Firefighting
- ESF #5, Emergency Management
- ESF #6, Mass Care, Housing, and Human Services
- ESF #7, Resource Support
- ESF #8, Public Health and Medical Services
- ESF #9, Urban Search and Rescue
- ESF #10, Oil and Hazardous Materials Response
- ESF #11, Agriculture and Natural Resources
- ESF #12, Energy
- ESF #13, Public Safety and Security
- ESF #14, Long-Term Community Recovery and Mitigation
- ESF #15, External Affairs

At this level in the United States, the most qualified and specialized units can be deployed in response to a disaster. These may include not only FEMA personnel but also other Department of Homeland Security (DHS) units, and extremely specialized teams located within other states. If a major earthquake were to take place in California, for example, it is extremely likely that the highly qualified urban search and rescue (USAR) team from Virginia would respond. The use of such vital resources is highly coordinated at a federal level.

Yet the most important contribution on the federal level is often in terms of policy. Actions by the Executive and Legislative branches can have significant and lasting impacts on the response phase. (See Chapter 2, Development and History of Emergency Management, for more information.) Such efforts often move in fits and starts, generally progressing only after a catastrophic event such as Hurricane Katrina or the September 11 attacks. Birkland notes that this systemic learning process is based upon an "appreciation of how problems come about, what causes them, and what can be done to change the conditions under which they grow worse" (2006, p. 50).

### Nongovernmental/Volunteer

Governments are not the only groups with a vested interest in effective emergency and disaster response. Multiple local, regional, national, and international organizations specialize in responding to crises in their particular area of expertise. Some of these groups are for-profit contractors, many specializing in disaster recovery operations for residents and businesses.

However, volunteer groups play an ever more-critical role in times of crisis. Volunteer Organizations Active in Disaster (VOAD) is a collective of dozens of groups possessing various response and recovery capabilities. Many of these individual organizations are faith-based, and see such volunteer work as an important piece of their community-outreach work.

Two of the largest and most capable groups active across the globe are the Salvation Army and the Red Cross. The Salvation Army, sometimes referred to as Salvationists, are a Christian-based organization that specializes in providing short-term amenities to disaster victims, such as food and clothing, as well as longer-term shelter. The International Red Cross/Red Crescent society has chapters across the planet and provides emergency rations and housing vouchers, among other services. Both groups are absolutely critical partners to emergency managers in cash-strapped urban agencies. (See Chapter 7, Collaboration in Urban Environments, for more information.)

Professional organizations can also be vital resources during a disaster. Organizations may focus on incredibly specific areas of expertise, such as art conservation or commercial pipefitting, and offer their members' expertise to aid in response and recovery efforts. Again, it behooves an urban emergency manager to have a very well-stocked contact list or Rolodex—both paper and electronic. Disasters are rarely predictable, and having a deep bench of subject-matter experts upon whom to call in a crisis is simply invaluable.

### Specialized Units

Finally, an emergency manager should have a complete knowledge of the highly specialized units at his or her disposal. Many state and local jurisdictions maintain incredibly well-trained specialists capable of responding to emergencies.

The aforementioned USAR teams are an excellent example. Requiring literally hundreds of hours of training to qualify, USAR team members can be deployed across the country—or internationally upon request—on just a few hours of notice. Generally designated as task forces within the ICS, such teams consist of a number of specific skillsets highly desirable in an urban disaster setting involving structural collapses or other types of entrapments (Fig. 12.3).

Other specialized teams are often located nearby as well. Large urban police departments commonly field tactical special weapons and tactics (SWAT) teams, capable of dealing with counterterrorism responses, and hostage/barricade situations. Disaster Medical Assistance Teams (DMATs) are medical professionals trained for extended field response. Urban fire departments are able to deploy hazardous materials specialists and technicians to respond to chemical releases or similar incidents, known as HazMat teams.

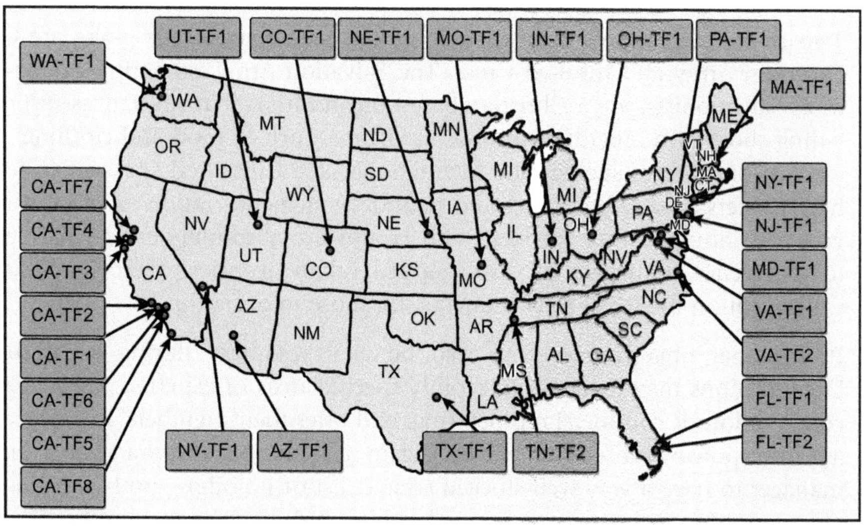


**FIGURE 12.3**

A map displaying the predeployment locales of the highly trained urban search and rescue (USAR) teams in the United States <https://www.fema.gov/task-force-locations>. *Source: FEMA.*

National Guard units often contain highly specialized squads with dual civilian/military applications, such as the Illinois National Guard's 5th Civil Support Team (CST). The unit specializes in weapons of mass destruction (WMD) scenarios but is also a highly competent resource in the event of an inadvertent large-scale chemical release or a transportation accident involving hazardous materials.

These specialized units represent just the tip of the iceberg in the available resources at the disposal of a well-connected emergency manager. Their skills are honed in hour upon hour of training and exercises, and they are almost universally eager and willing to put such knowledge to practical, operational use.

## Mass Notification

Another critical piece of an emergency manager's arsenal is the ability to communicate widely in a very compressed timeframe. Such capability is referred to as mass messaging or mass notification. In an increasingly connected world, civilians expect to hear definitive and authoritative messaging coming from public-safety officials in times of crisis.

The traditional use of mass media (primarily television and radio) is being rapidly outstripped by the digital age. In the world of Twitter, Instagram, and WhatsApp, such communications channels seem slow and obsolete. The unfortunate fact is that such applications and programs are counterproductive during the initial response phases, as they will be almost entirely filled with rumor and innuendo, but very little fact.

Public-safety officials, and emergency managers in particular, must be both authoritative and prompt in their messaging. Emergency managers in corporate and nonprofit environments face similar challenges and expectations.

There are two key components to such a program being successful. The first is a capable and flexible platform, and the second is the right balance between the need to be right and the need to be first.

An urban jurisdiction should make every effort to establish connections with its citizens through multiple channels. Reverse-911 is a solid technology based on sending out a prerecorded message to every landline phone in a given geographic area, and should not be overlooked. Yet with the migration to mobile phones and e-mail, jurisdictions must make a concerted effort to be able to send texts and e-mails to reach people in their homes, at their workplaces, and on the move. This requires partnerships with cellular and wireless providers, and likely a contract with a messaging system provider. The platform must reach as many people as quickly as possible during an emergency, and thus must be secure, relatively easy to use, and capable of very high volumes.

The second aspect of a successful mass-messaging program is balance. The fact is that everyone with a smartphone, laptop, or tablet device will start sending out muddled and contradictory information if they are near a large-scale incident. Public-safety officials are never going to have their message out first, so that message needs to be highly authoritative and reliable when it is released. That said, a long delay in releasing any official statement creates a vacuum in which rumor and false information can spread rapidly, creating undue panic.

A balanced approach is therefore best, releasing key basic facts early in the response phase of a large-scale incident, and then shifting to more detailed information as it becomes available and proper to release. TV, radio, newspapers, and news websites become completely appropriate mediums to utilize as the response phase moves into recovery. A joint information center (JIC) is highly valuable for any multijurisdictional incident, as is recommended by the ICS. (See Chapter 7, Collaboration in Urban Environments, for more information.)

Finally, it is absolutely critical that there be a preestablished protocol and access list for who is specifically authorized to make statements on behalf of a given jurisdiction or organization. We have all witnessed this process be handled extremely well and be handled extremely poorly—often by the same organizations. This function will have an outsized impact on the perceptions of stakeholders long after the response phase is concluded. The difference is, once again, having a plan *before* the crisis occurs.

## Training and Education

Training and expertise have been key factors in earlier descriptions of some of the most specialized units in the emergency-management realm. The aforementioned USAR teams and SWAT teams, for example, spend a majority of their time in training and exercise mode—striving to make their actions and procedures "muscle memory" for an actual incident.

Emergency managers themselves must actively and aggressively emulate this behavior. Training and education are an invaluable part of what a qualified emergency manager brings along in a response situation. Whether in an EOC or at a forward command post, fellow public-safety professionals expect a certain level of professional expertise from their cohorts. (Chapter 2, Development and History of Emergency Management, and Chapter 10, Preparedness Phase, each contain additional guidance.) Two key forms of training can have outsized influence on the response phase.

One of the most important activities in which an emergency manager can participate is drilling and exercising. These activities offer the closest group simulation that can be effectively presented prior to an actual response. Emergency-management practitioners should take every opportunity available to partake in drills and exercises and to be involved in their planning and execution whenever possible. Coaches are absolutely correct when they state that you practice how you play. And the more practice you have before "game day," the better mental shape you'll be in when the time comes.

The second learning aspect closely related to response is often less formal. A hotwash is an after-action review taken immediately following an actual incident or emergency. It generally consists of a broad group of responders sitting around, often still in their field gear, and recounting what they have just been through while it is still fresh. This represents an excellent opportunity to collect raw data and accurate memories before they begin to fade with time. It may be a challenge to get responders to sit down for just a few minutes when they are so close to being dismissed from an incident scene—but it is time well spent that may save a life on the next shift or at the next large-scale response.

### Transitioning From Response to Recovery

Finally, a note on the shift from a response operation to a recovery operation. The fact is that this transition is rarely a bright, clear line during an operation. Instead, it is generally a gradual pivot from primarily response tactics to primarily recovery tactics.

It cannot be overstated that this must be handled with great delicacy if victims are still missing or unaccounted for. Always presume that you are being observed by the media and the public. Family and friends of the victim may interpret the term "recovery" to mean that first responders are no longer looking for their loved one. The very use of the word conjures an element of finality for some.

However, from an operational standpoint, this is an important transition that must be made in a timely manner in order to begin restoring a sense of normalcy and to begin preparing for the next major crisis that may emerge. This shift will also represent a time of reflection, and a time to consider lessons learned that can be applied to the next deployment.

## CONCLUSIONS AND IMPLICATIONS

The response phase represents the most chaotic and dangerous portion of a major emergency. A brief overview of some of the hard-earned lessons of responding to emergencies in urban settings is worthwhile.

## Information Versus Rumor

It has been said that 50% of initial reports on a major emergency will be incorrect or inaccurate. And that percentage may be kind. As an emergency manager in an urban setting, presume that much of what you hear in the first chaotic moments of a large-scale incident is flawed. Until you have trusted eyes and ears at the scene reporting to you at the EOC or forward command post, take every word with a grain of salt.

## Be a Generalist—And a Specialist

Emergency management is a highly unique profession in that practitioners are expected to have a broad general knowledge of the field, and also to bring something unique to a response. It is critical that an effective emergency manager have a basic understanding of his or her fellow responders in law enforcement, fire services, and emergency medicine. The true stars in the field also possess a specific area of subject-matter expertise that is both relevant and runs deep.

## Presume It Is Terrorism

When you arrive at an incident scene, or take your seat in an EOC, presume that every single incident is terror-related until proven otherwise. Particularly in a field-response role, emergency managers must be cognizant that terrorists often target first responders with secondary devices, or follow-up attacks. Don't presume that the threat has passed merely because a lot of vehicles with flashing lights have arrived on scene.

## Be a Team Player

A lot of first responders have enormous egos—there, the truth is out. The emergency services tend to attract Type-A personalities with a lot of confidence. Disagreements will result, particularly during stressful responses. As an emergency manager, you may be seen by others as a "referee" or neutral party. Don't hesitate to use that fact to your advantage during the course of a response.

## Assume That You Are on Camera

In an age where smartphones and digital cameras are everywhere, several will be present at any major response scene. Responders have always had a sense of gallows humor, but a group of uniform personnel briefly standing around and sharing a lighthearted moment or joke can be interpreted as callous or disrespectful behavior. Remember that images can be taken out of context, and perceptions matter.

### Take Care of Yourself

It is easy to get so focused on a response that you forget to look out for Number 1—yourself. You won't be coming to anyone's rescue if you forget to eat, hydrate yourself, or get adequate rest. Most large-scale responses are marathons, not sprints. Whether you are in the field or in an EOC many miles away, listen to what your body is telling you. There is a reason that airlines announce to "put your oxygen mask on first before assisting others" during an emergency.

## SUMMARY

A passenger train collision in Chicago serves as a real-world example of emergency management during the response phase. Urban emergency managers should be prepared to respond to crises with natural, human-caused, or hybrid causations. Planning will be critical to any effective response.

Large-scale responses may involve resources from local, state, regional, federal, and nonprofit entities as well as highly specialized units. Modern emergency managers may respond to incidents in the field, or help manage outcomes from emergency operations centers or similar facilities. Mass notification, and prior training and exercising, will have a significant impact on the outcome of a given response.

### References

Bevelacqua, A., & Stilp, R. (2009). *Terrorism handbook for operational responders*. Clifton Park, NY: Delmar.

Birkland, T. (2006). *Lessons of disaster: Policy change after catastrophic events*. Washington, DC: Georgetown University Press.

Erickson, P. (2006). *Emergency response planning*. Burlington, MA: Butterworth-Heinemann.

Sylves, R. (2008). *Disaster policy & politics*. Washington, DC: CQ Press.

### Further Reading

Haddow, G., Bullock, J., & Coppola, D. (2008). *Introduction to emergency management*. Burlington, MA: Butterworth-Heinemann.

National Transportation Safety Board. (2009). Retrieved from: <https://app.ntsb.gov/investigations/summary/RAR0901.html>.

# CHAPTER 13

# Recovery Phase

## INTRODUCTION

By late September of 2001, the dust had largely settled and the response phase was concluding across the island of lower Manhattan. Two weeks earlier, this had been one of the sites of the worst acts of terrorism in American history, killing well over 2000 people and causing in excess of $3 billion dollars in damages at the World Trade Center location alone.

It had not been a good fortnight, even for the survivors. The national mood was one of shock and anger. Funerals for both civilians and first responders continued to take place on a daily basis. Ten of thousands remained temporarily displaced and essentially homeless in the short term. Tens of thousands more had lost their jobs when more than 12 million square feet of office space was damaged or destroyed. The stock markets had taken a mighty loss in the days that followed the attacks. Entire sectors of critical infrastructure were heavily damaged or partially crippled, and public transportation was snarled.

And every day and night, first responders continued to dig for the bodies of victims. A cadre of volunteers worked in an amazing array of support roles, from food preparation and first aid to equipment maintenance and grief counseling. Occasionally, all work would cease and responders would fall silent, remove their helmets, and watch the remains of a fellow public-safety professional be pulled from the wreckage and carried to a waiting ambulance. And then they would turn around and start digging again. New and unwelcome terms had been added to the lexicon of New Yorkers—the pile, a new normal, ground zero. Wild conspiracy theories began to circulate, growing ever more complex and convoluted.

It was, and remains, one of the most visible and painful examples of the recovery phase following a disaster in modern history. And the true rebuilding would continue for many, many years to come.



Urban Emergency Management. DOI: http://dx.doi.org/10.1016/B978-0-12-805307-2.00013-5

## ANALYSIS

As the response phase winds down, the recovery phase takes over. As with all phases, there is often no clear signpost as one ends and the next begins. For example, even as an industrial accident cleanup is completed, or a contract agreement is reached to end an acrimonious employee strike, a company should be enacting the recovery steps of its continuity of operations plan. Even as the last search and rescue teams leave a disaster scene, the jurisdiction must be activating its recovery protocols. In essence, an organization cannot wait for some divine sign that the response phase has been completed prior to initiating recovery efforts.

The recovery phase is also somewhat more challenging to define than the other phases, noted by Haddow, Bullock, and Coppola. The function begins in the initial hours and days following a disaster and can continue for months and, in some cases years, depending on the severity of the event (2008, p. 155). Recovery is the process of getting back to "normal"—even if it is a new and different sort of normalcy.

Recovery may involve an incredibly wide array of tasks and stakeholders. From a tangible reconstruction perspective, a single disaster may necessitate clearing debris, restoring utility service, rebuilding homes, rebuilding businesses, replacing property, and hardening infrastructure. Simultaneously, the mass-care elements may include providing temporary housing, clothing, food, and counseling services. While this is obviously a highly complex operation, the phase of recovery is capable of dovetailing with mitigation and preparedness to leave a community or organization stronger after a disaster occurs than it had been previously.

For an organization, it is vital to note that recovery will depend in great part on the previous phases of preparedness, mitigation, and response. (Recovery presumes that the incident has indeed taken place, so the prevention phase did not fully develop.) If each of these phases has gone poorly or been neglected, then recovery is likely to be a longer and more complicated process.

### Plan for a Return

As in the other phases, planning will be key—particularly in areas such as finance, zoning, and housing. Recovery processes will vary for any crisis, but may include elements of rebuilding homes and commercial facilities, replacing property, resuming employment, and reputational management.

Therefore, as a part of the preparedness process, an organization must ready itself for any eventuality along that spectrum, from a minor interruption of normal operations to a catastrophic impact that requires significant and long-term changes. This is generally referred to a contingency planning, making an organization more flexible under stress. Thus, in many ways the cycle completes itself, with recovery feeding back into the phases of prevention and preparedness.

The professional security association ASIS International releases a series of best practices and guidelines, including the *Business Continuity Guideline* (2004). Here are the sample planning questions the organization recommends be asked as a part of continuity planning:

- Has your organization planned for survival?
- Is the business continuity plan up to date?
- Have the internal audit, security, and insurance units reviewed the plan?
- Has a planning team been appointed?
- Has a risk assessment been conducted?
- Are people protected?
- Have critical business processes been identified and ranked?
- Is technology (i.e., data, systems) protected?
- Have resources required for recovery been identified?
- Have personnel been trained and exercises conducted?
- Has a crisis-management center been identified?
- Have alternative worksites been identified?

While this checklist was written from a private-sector viewpoint, its core tenants are clearly applicable to both the public and nonprofit sectors. From a continuity standpoint, there is significant commonality across organizations. Just as a threat and hazard assessment will inform the prior phases of emergency management, the recovery plan must take into consideration all possible impacts and outcomes that may result from said threats.

## Resiliency Pays

Resilience and resiliency are crucial concepts within the field of disaster science and have an immense impact on the phases of preparedness and mitigation that should take place before a disaster or major disruption. As addressed in Chapter 10, Preparedness Phase, resiliency refers to making a place or thing more resistant to damage or disruption.

This concept will also inevitably come into play during the response and recovery phases. The National Infrastructure Protection Plan (NIPP) defines resiliency as the capability of an asset, system, or network to maintain its function during an emergency or disruption. No system or network is completely impervious to damage or disruption, and it is thus worth acknowledging that shutting down to prevent catastrophic damage, or failing in a predictable or somewhat controlled fashion, may also be a worthy objective.

An emerging example of resiliency planning comes in the form of the 100 Resilient Cities (100RC) program. Funded in large part by the philanthropy of the Rockefeller Foundation, this international effort seeks to address both chronic and urgent or acute needs in large metropolitan areas. Assembled between 2013 and 2016, the list of participating cities had to meet a number of qualification criteria, including the participation of willing and innovative officials. The organization views resilience as not just the immediate recovery phase, but rather planning for and addressing "the stresses that weaken the fabric of a city on a day-to-day or cyclical basis." Participating cities receive financial and logistical assistance to name a Chief Resilience Officer for their jurisdiction and are also plugged into a standing network of subject-matter experts and planning professionals.

Indeed, the concept of resilience has become widely accepted in fields as varied as structural engineering and sociology, and multiple guidance documents have been produced in recent years. From a disaster sciences perspective, the Australian Emergency Management Institute (2011, p. 30–31) lists the characteristics of a resilient community or organization as:

- Trust
- Social cohesion
- Supporting attitudes and values
- Leadership
- A sense of community
- Good communication and information
- Collective efficacy
- Community involvement
- Resource dependency
- Social capital
- Existing norms
- Engagement with government

And while this checklist was written from a public-sector viewpoint, its core tenants would apply to corporate or nonprofit entities as well. Once again, the commonalities across sectors are striking.

## Causation Impacts

The root cause of a given emergency or disaster will clearly have an immense impact on the scope, scale, and nature of the recovery process. Each of the major hazard types has certain unique characteristics or qualities that will affect the response phase.

### *Natural Hazards*

Natural hazards are the most common cause of recovery efforts. Flooding, in particular, generates an immense amount of damage and displacement to urban jurisdictions and organizations.

As a broad range of hazards, natural causes may offer some warning of their approach (severe storms, flooding) or none at all (earthquakes). The commonality is widespread damage of varying degrees. Flooding is likely to be more severe near a body of water and less severe inland. A tornado may remain on the ground for several miles, but the severe damage will be focused along its relatively narrow track.

Natural hazards also tend to be recurring in nature. If a certain subdivision of houses flooded during a period of heavy rain, the incident is likely to recur the next time an equally heavy rainfall takes place. If an earthquake takes place that is severe enough to cause structural damage, it is likely that future tectonic movements will take place in the same region. Data regarding this likelihood of recurrence, even if rare, will inform long-term recovery efforts {{{Table 13.1}}}.

### *Human-Caused Hazards*

Human-caused or manmade hazards are less predictable. They will likely take place with little or no warning. Public-safety officials rarely have specific details and timeframes for pending technological failures or terrorist attacks.

The resulting impacts tend to see damage focused upon one geographical area or system. A bridge failure will be a large-scale but geographically

**Table 13.1** Estimated Recovery-Phase Expenses for Costliest US Hurricanes

| Storm Name | Year of Landfall | Estimated Cost |
|---|---|---|
| **1.** Katrina | 2005 | $47.4 billion |
| **2.** Andrew | 1992 | $23.3 billion |
| **3.** Sandy | 2012 | $18.8 billion |
| **4.** Ike | 2008 | $13.3 billion |
| **5.** Wilma | 2005 | $11.9 billion |
| **6.** Charley | 2004 | $8.9 billion |
| **7.** Ivan | 2005 | $8.5 billion |
| **8.** Hugo | 1989 | $7.0 billion |
| **9.** Rita | 2005 | $6.5 billion |
| **10.** Francis | 2004 | $5.5 billion |

*Source: CNBC.*

specific incident. Cybercrime may impact many individuals, yet is generally concentrated around a single system vulnerability or database.

Adding to their complexity, human-caused hazards may or may not be recurring in nature. If a bridge collapses, it will not do so again. However, if there are structures with largely identical designs in operation elsewhere, they must be considered as a potential risk for a similar failure.

### *Hybrid Hazards*

Hybrid hazards represent a wildcard in terms of impacts. As they are, by nature, cascading incidents with multiple inputs and variables, thus they represent a true range of geographic hazards.

The previously cited Fukushima disaster is a nearly worst-case example, combining essentially three hazards (earthquake, tsunami, nuclear accident) into a single cascading failure. While emergency managers may have been prepared for one or even two of these hazards to impact their jurisdiction, it is not unreasonable to state that any set of checks and balances would have been overwhelmed by this series of crises.

Because of this "falling dominoes" effect, hybrid hazards have a capacity to be geographically large and are almost inevitably more complex and multilayered recovery operations than natural or manmade hazards alone.

## The FEMA Model

The US federal government, in fits and starts, has been developing a set of guiding principles for disaster recovery over the past several decades. The Federal Emergency Management Agency (FEMA), now a part of the larger Department of Homeland Security (DHS), is the government's lead agency for all phases of emergency management.

FEMA's capabilities and focus have ebbed and flowed based on a number of factors, including leadership capability and Congressional funding. (See Chapter 2, Development and History of Emergency Management, for more information.) It would be ideal for the agency to be viewed as a nonpolitical entity, but in the current acrimonious environment in Washington, D.C., this seems highly unlikely.

The James Lee Witt era of the 1990s saw a refocusing of the agency on mitigation and recovery operations. Much of that structure remains in place and has been recently reinvigorated by recent agency director Craig Fugate, himself a former emergency manager in the state of Florida.

One of the traits that Witt and Fugate held in common was a focus on minimizing both current and future impacts and costs associated with disasters.

Both held an all-hazards viewpoint, prioritizing programs and projects that could help address multiple types of risks.

Over the years, this has resulted in a FEMA that addresses all phases of emergency management, and also acts as a funding mechanism for state and local entities. Multiple FEMA grant programs provide jurisdictions and organizations with "motivation money" to improve the status of their emergency planning, critical infrastructure, or public outreach programs.

And yet the largest and most variable piece of FEMA funding applies directly to recovery efforts. To assist a jurisdiction or a critical organization get back on its feet following a disaster, FEMA provides a range of financial incentives to drive recovery efforts. These include block grants to state and local governments, individual grants for temporary housing and housing repairs, and low-interest loans to impacted businesses.

The federal government also makes technical assistance and subject-matter experts available to qualifying entities. Without these funds and expertise, designed to jumpstart local recovery efforts, disruptions would unquestionably be longer and more severe following major emergencies.

## Other Federal Agencies

FEMA does not, however, stand alone when it comes to disaster recovery. Once a given incident has been declared an emergency or disaster by the rules of the Stafford Act, multiple federal resources become available to victims and their communities. (See Chapter 2, Development and History of Emergency Management, for more information.) The assistance offered by various agencies can be quite specific to their area of expertise, but taken as a whole, it offers emergency managers in the United States a wide range of potential options.

### *Department of Agriculture*

Low-interest loans may be made available to farmers, ranchers, and other producers impacted by a disaster. The loans can be used to replace critical machinery or facilities, pay production costs, or refinance debt.

### *Department of Commerce*

The Economic Development Administration provides economic assistance for communities directly impacted by a presidentially declared disaster, when properly funded by Congress. The EDA provides grants for infrastructure development and unemployment assistance in areas negatively impacted economically by disaster.

### *Department of Health and Human Services*

The DHHS serves as the lead federal agency for health and medical issues following a declared disaster. It provides a range of services often focused on mental health and counseling as well as assistance and programs for the elderly.

### *Department of Housing and Urban Development*

Department of Housing and Urban Development (HUD) provides housing grants following disaster, particularly for low-income neighborhoods adversely impacted. It also oversees the crucial Community Development Block Grant program, which provides seed money and financial incentives to accelerate the recovery process in urban areas.

### *Department of Labor*

The agency oversees the Disaster Unemployment Assistance Program, which provides bridge funds to individuals whose employment has been directly affected by a declared disaster. The Department of Labor may also be authorized to hire displaced workers to serve as a temporary workforce to assist with the recovery effort.

### *Department of Transportation*

The agency oversees the Emergency Relief Program, which supplements states in paying for critical transportation—mostly road and rail—repairs following disaster. Department of Transportation (DoT) may also seek accelerated replacement or repair of crucial federal assets such as interstate highways, although Congressional funding is always a challenge.

### *Small Business Administration*

The SBA is one of the more active players among federal agencies following a disaster. Its primary role is to offer very low-interest loans to small businesses affected by a declared disaster in order to accelerate the recovery process. Approximately $30 billion of these loans have been granted since the program's inception, the vast majority of which have been successfully repaid to the US Treasury over time.

### *US Army Corps of Engineers*

The Corps of Engineers offers a wide range of engineering and public works expertise and responds to more than two dozen declared disasters in an average year. If authorized by their command structure, they can also be utilized in state-level emergencies.

## The Nonprofit Model

Among the many vital stakeholders across the phases of emergency management, volunteer organizations play an absolutely vital role. In the United States in particular, these groups are not considered as unnecessary or superfluous, but rather as key components of the response and recovery processes.

As federal, state, and local capabilities vary widely depending on administrations, competencies, and fiscal funding, the need for a safety net of qualified volunteers has become crystal clear. The National Voluntary Organizations Active in Disaster (National VOAD) is essentially an association of associations. Its members, most with religious affiliation, are trained in a range of response and recovery procedures and have proven invaluable in large-scale disasters. As of 2016, there were 58 active organizations within NVOAD:

- Adventist Community Service
- Adventist Disaster Response
- Alliance of Information and Referral Systems
- All Hands Volunteers
- Amateur Radio Emergency Service
- American Red Cross
- Billy Graham Rapid Response Team
- Brethren Disaster Ministries
- Buddhist Tzu Chi
- Catholic Charities USA
- Churches of Scientology Disaster Response
- Church World Service
- Convoy of Hope
- Cooperative Baptist Fellowship
- Christian Church Disciples of Christ
- Direct Relief
- Episcopal Relief & Development
- Feed the Children
- Feeding America
- Habitat for Humanity
- Headwaters
- Heart to Heart International
- Hope Animal-Assisted Crisis Response
- Hope Coalition America
- Hopeforce International
- Hope Worldwide
- The Humane Society

- ICNA Relief USA
- Institute for Congregational Trauma and Growth
- International Critical Incident Stress Foundation
- International Orthodox Christian Charities
- Islamic Relief
- The Jewish Federation of North America
- Latter-Day Saint Charities
- Link 2 Health Solutions
- Lutheran Disaster Response
- Mennonite Disaster Service
- Mission to North America
- National Association of Jewish Chaplains
- National Organization for Victim Assistance
- Nazarene Compassionate Ministries
- Nechama
- Operation Blessing
- Partnership with Native Americans
- Points of Light Institute
- Presbyterian Disaster Assistance
- Rebuilding Together
- Samaritan's Purse
- Save the Children
- Society of St. Vincent de Paul
- Southern Baptist Convention Disaster Relief
- The Salvation Army
- Team Rubicon
- Toolbank Disaster Services
- United Methodist Committee on Relief
- United Church of Christ Disaster Ministries
- United Way
- World Renew Disaster Response Services

## Revisiting Katrina Recovery

As maligned as the federal response to Hurricane Katrina has been, FEMA and its fellow agencies unquestionably learned some valuable lessons from both successes and mistakes during the recovery phase. The storm displaced more than 750,000 residents, and more than 30,000 were rescued from imminent danger by DHS assets (primarily the US Coast Guard).

Critical infrastructure such as power, levees, and schools were badly damaged during the storm, and by the standing water that followed. According to the federal Executive Branch report that followed, housing damages alone totaled

$67 billion, with business properties hit for $20 billion and government facilities at $3 billion. The total damage estimate across all phases ranges as high as $108 billion, making Katrina by far the most expensive disaster in US history.

Beyond the financial cost was a grave erosion of confidence in all levels of government—and particularly the feds—to protect their citizens. (New Orleans Mayor Ray Nagin was soon convicted on corruption-related charges based in part upon dishonest dealings during the hurricane recovery.) The response had been slow and disorganized, particularly from the perspective of those most in need of assistance. As the transition from response to recovery got underway, progress continued to be hampered by the same infighting and lack of coordination. {{{Fig. 13.1}}}

Haddow et al. referred to early recovery efforts as ineffectively designed and implemented due to issues of leadership, coordination, and planning (2008, p. 402). In addition to the vast sums of money spent by FEMA and other federal agencies, corporations and foundations contributed over $1 billion dollars to specific causes and efforts within the larger recovery. The variance in governance and in funding sources has led to a very uneven recovery for the region as a whole.

A decade after the hurricane, the population of New Orleans remains more than 100,000 less than it had been according to the 1990 census. Many Gulf Coast residents will simply not be returning to the area, and "expat" communities of Katrina refugees have established themselves in neighborhoods from Atlanta to Houston.

**FIGURE 13.1**

The raw power of hurricanes is fully displayed by the destroyed vehicles and structures they leave behind, in this case Hurricane Katrina in 2005. *Credit: Data from CNBC.*

## A Better Way

If the Hurricane Katrina recovery effort left much to be desired, what lessons can emergency managers take away from the experience? What factors constitute a successful recovery? Some consensus is beginning to form around these questions. The assessments of the federal response and recovery efforts for Hurricane Sandy, less than a decade later, were universally more positive. While not a checklist for every hazard or every eventuality, it is clear that a number of key, recurrent challenges must be addressed during recovery operations:

- Immediate Safety (Distance from Threat)
- Immediate Shelter
- Food/Rations
- Medical Care/First Aid
- Chain of Command
- Situational Awareness
- Security/Law Enforcement
- Transportation
- Debris Removal
- Utility Restoration
- Mass Notification
- Logistics Networks

Several key themes emerge from the multitude of after-action reports following modern large-scale disasters, including Hurricane Katrina. Each has a large degree of bearing and relevance to the recovery phase and is summarized below.

### *Leadership Matters*

In times of crisis, both responders and the civilian population will seek leadership and guidance. The power vacuum in response and recovery efforts along the Gulf Coast in 2005 was tangible. By contrast, New York Mayor Rudy Giuliani displayed many of his faults following the September 11 attacks, yet was a highly visible presence within the city. General Russel Honore took on much the same role in the Katrina recovery, and again in the Deepwater Horizon oil spill of 2010. Establishing clear leaders early in an incident is crucial.

### *Follow Maslow*

Maslow's famous treatise on the hierarchy of needs still holds true today. The immediate priorities in a true disaster should be safety, food, and shelter. All else is secondary. If you are able to provide a sense of safety to victims, and a clear path to how they will achieve shelter and food, you have won the short-term battle. All other needs become secondary when in crisis mode.

### *Communicate*

Every after-action report in human history has included some variant of the phrase "communication could have been better." Communication should not focus just on radios and cell phones, but on who exactly is in the loop. Officials should be communicating both horizontally between jurisdictions or departments, and vertically between levels of authority. Situational awareness is absolutely vital to successful response and recovery efforts.

### *Prepare and Mitigate*

As addressed at the beginning of this chapter, preparedness and mitigation efforts will be vital during the response and recovery phases. If they are ignored or underaddressed prior to a crisis, there will be no opportunity to correct that oversight until well into the recovery phase. Ultimately, such efforts reduce the overall impact of a given hazard, thus making it a shorter path back to normalcy during the recovery phase.

### *Flexible Planning*

No two disasters will unfold precisely the same way. Writing broad, realistic plans that can be adapted to wide ranges or categories of related risks or hazards should be a top goal of any organization or jurisdiction. Plans must be multijurisdictional, and drilling or exercising with multiple agencies or departments is highly desirable.

### *Rumor Loves an Information Vacuum*

Providing valuable updates to victims can allay fears and stop the rumor mill in its tracks. Setting a schedule for media briefings, or in-person updates in shelters and similar locations, is invaluable. If new information is not available, keep the pipeline open by stating so, and providing the time and location of your next update.

### *Coordinate*

No single agency or organization is capable of handling a major emergency completely independently. Part of the planning process is knowing which partners have which resources. Coordination during the response and recovery phases must include not just areas of responsibility but also matters of resource allocation and logistics.

## CASE STUDY: THE GREAT MIDWEST FLOOD OF 1993

### The Facts

More than a decade before Hurricane Katrina struck the Mississippi River delta region, another deluge took place a few hundred miles upstream. In the spring, summer, and fall of 1993, the Midwest United States received an extremely high amount of rainfall. Melting snowpack from the eastern Rockies exacerbated the problem. A number of major rivers, including the Mississippi and the Missouri, reached and then exceeded their flood stages. And the water just kept coming, through June, July, and into August.

Disaster declaration zones were established across nine Midwestern states, including 532 counties and—unprecedentedly—the entire state of Iowa. The volume of water grew so large that major levees began to fail and entire towns were in danger of being inundated. TV viewers at the time witnessed the surreal image of a virtually intact, two-story white farmhouse floating down the swollen channel of the Mississippi River. The losses due to the flooding would eventually total more than $12 billion. When the waters finally began to recede back within their banks that fall, a thick coating of mud and sludge covered every surface they had swallowed during the summer months.

### Analysis and Investigation

The damage from the massive and widespread flooding was severe, and the recovery process would be a lengthy one. The unholy mess left behind served as a pungent reminder of that fact.

St. Charles, Missouri, provides an example of some of the hardest hit jurisdictions. The county, and city of the same name, lay at the exact juncture of the Missouri and Mississippi Rivers. This made for extremely fertile farmland, led to the rapid growth of the area in the 1800s, and even made it the perfect point for the crew to embark upon the famous Lewis and Clark Expedition of 1804. In 1992, it was the fastest growing county in the state by population.

Yet the massive river juncture had its drawbacks as well. Flooding was not uncommon, and the jurisdiction had both a system of large levees, and areas of floodplain set aside as wetlands. Throughout much of its history, St. Charles had benefited greatly from being located at the intersection of the two longest rivers in the United States. That was definitively not to be the case in 1993.

At the crest of the flooding, fully 30% of St. Charles County was underwater in 1993 (Sylves, 2008, p. 122). The levee system protecting the wedge-shaped county held throughout the spring, but began to become inundated and overtopped as the water kept coming during the summer months. One by one, the levees began to fail. Acre after acre of farmland disappeared beneath the murky, oily waters. Then 2000 homes and thousands of other structures did as well. Damage estimates for the county—a mix of urban, suburban, and rural areas—exceeded $150 million.

As the waters finally began to recede, the recovery process could truly get underway via damage assessments and inspections. Simultaneous debris removal and utility restoration continued at a breakneck pace. Structures suffering damages totaling more than 50% of their assessed value were condemned. Other less-damaged properties were required to be rebuilt at greater base heights—using stilts or berms—if they were to receive funding for reconstruction. County officials served as the ultimate gatekeeper to issuing permits for demolition or reconstruction, but followed the guidelines established by the federal National Flood Insurance Program (NFIP).

This process made for some very tense discussions and decisions among property owners, but ultimately resulted in a significantly lower number of properties—particularly residences—being located in flood-prone, low-lying areas near the two rivers. Federal officials made themselves available by setting up local offices, and holding multiple town halls and educational sessions to explain the process. Not every property owner was satisfied, but the vast majority worked within the process to achieve acceptable results.

What makes the area particularly salient as a case study is that another flood, slightly smaller but nonetheless severe, stuck the exact same region in 1995. The property-damage claims in St. Charles were a fraction of the incident 2 years prior, and many critics finally warmed to the concept of property buyouts and zoning restrictions.

The full arsenal of options utilized to achieve these results is worthy of review. Haddow et al. (2008, p. 170–171) make note of several effective and valuable mitigation and recovery planning tools brought to bear in the wake of the 1993 floods:

- Land-Use Planning Techniques: Including acquisition, easements, annexation, stormwater management, and environmental reviews
- Zoning: Including permitting, historic preservations, setbacks, density controls, wetland protection, floodplain and coastal management
- Building Codes: Including design control and review, height and type regulations, and soil stability ratings
- Financial Incentives: Including special districts, tax exemptions and incentives, buyouts, special bonds, development rights, property transfer, or use change fees
- Information and Oversight: Including public awareness and education, regional planning agreements, and public meetings

### Case Study Conclusions

Ultimately, the recovery phase of the Great Midwest Flood of 1993 was objectively and subjectively more successful than the same phase following Hurricane Katrina 12 years later. Why was this the case?

The two disasters bore many key similarities. Both affected multiple states across a single geographical region. Both came with some warning that allowed for a degree of planning and preparation. Both would leave behind widespread damage and debris which would require significant federal support to remedy.

Yet there were important differences to consider as well. The population density of many areas impacted by Katrina was significantly higher. Many coastal residents depended upon public transportation, while Midwesterners largely did not. The Midwestern floodwaters were a visible, rising threat, while the hurricane's predicted severity was discounted by many until it was too late.

The City of St. Louis was able to close the massive flood gates at the base of the Gateway Arch along the Mississippi River, and unlike the levees in New Orleans, they held. And thus the densest urban area directly impacted in 1993 was largely a dry island of land surrounded by floodwaters, while New Orleans had no such luxury in 2005.

Finally, the Witt-inspired mating of recovery and mitigation procedures was immensely successful. By utilizing a combination of land-use planning, zoning, building codes, financial incentives, and public information, the response phase to the Great Midwest Floods achieved every major milestone. In assisting victims to recover from one flooding event while simultaneously minimizing the impact of similar future events, local, state, and federal governments made very good use of planning tools and of taxpayer funds.

## CONCLUSIONS AND IMPLICATIONS

The recovery phase is almost universally the longest and most expensive part of a major incident. As an emergency manager, it can also be one of the most frustrating and exhausting. And yet there are a few tips that can help compress both the timeframe and the frustration.

### Document Everything

The recovery process can be a slog. Days will blur together, and months later specific memories will be hard to come by. So write it all down. Use incident command forms where appropriate, but do not hesitate to use notebooks or spreadsheets to record basic data or facts. Your Administrative/Financial Section Chief will thank you when federal agencies want documentation of your time and expenses.

### Establish a System Early

Much of your time during the initial portion of the recovery phase will likely be spent performing inspections and damage assessments. Before you set out on your first such mission, sit down with your fellow assessors and establish some basic ground rules. Set strict criteria for what constitutes light, moderate, and severe damage. The home or facility you visit on day 60 deserves to be measured by the same standards as those you visited on day 1.

### Communicate With Victims

One of the most common complaints heard following a large emergency or disaster is that authorities failed to communicate with the public. Have a few go-to pieces of information that you can offer to people impacted by a major incident. Offering points of communication such as field offices or toll-free phone numbers is often welcome, as is having a general outline of next steps for victims.

### Don't Overpromise

Recovery takes time—probably more time than you first anticipate. Residents and business owners may press emergency managers for commitments that everything will be "back to normal" next week or next month. Be honest. Lay out the complex challenges in the recovery process, and ask victims to assist in getting things back on track in areas they can control.

## SUMMARY

The phase of recovery may be difficult to clearly delineate, and often represents the most costly and lengthy of the phases of a disaster. Planning, causation, and resiliency will all be important factors after an incident has taken place. A number of previous after-action reports offer guidance and best practices for the recovery phase.

FEMA and supporting federal agencies each have roles to play during recovery. The Great Midwest Floods of 1993 provide a worthy case study of a relatively successful recovery process and offers several potential lessons for future large-scale incidents.

## References


ASIS International (2004). *Business continuity guideline.* Alexandria, VA: ASIS.

Australian Emergency Management Institute. (2011). Retrieved from: <https://www.ag.gov.au/EmergencyManagement/Respond-and-recover/Documents/aemhs-2-community-recovery.pdf>. Accessed on June 2016.

Haddow, G., Bullock, J., & Coppola, D. (2008). *Introduction to emergency management.* Burlington, MA: Butterworth-Heinemann.

Sylves, R. (2008). *Disaster policy & politics.* Washington, DC: CQ Press.


## Further Reading


100 Resilient Cities. (2016). Retrieved from: <http://www.100resilientcities.org>. Accessed on January 2017.

Department of Homeland Security (2008). *National infrastructure protection plan.* Washington, DC: U.S. Federal Government.

Roberts, P. (2006). FEMA after Katrina. *Policy Review.*

White House (2006). *Federal response to Hurricane Katrina: Lessons learned.* Washington, DC: U.S. Federal Government.

CHAPTER 14

# Future of Urban Emergency Management

## INTRODUCTION

Predicting the future can be a notoriously risky proposition. Making such predictions all the more perilous is the fact that emergency management is both a young profession and a young field of study. With less than half-a-century of time under its belt, emergency management is still in its formative stages when compared to established areas of expertise such as sociology, or fire science, or logistics.

Yet what it has in common with these seemingly disparate fields is that it is both an academic *and* operational pursuit. On the academic side, research and analysis continues to accrue regarding the basis and foundations of the science of risk reduction, and effectively managing emergencies and crises. On the operational side, practitioners continue to gather valuable field experience and best practices to pass along to the next generation.

Both internal and external forces will impact the profession in the immediate future. It is, in short, a very exciting and rewarding time to be an emergency manager. It is an optimal time to look back and to look forward.

## ANALYSIS

### Back to the Beginning

As a function, emergency management has indeed been with us a very long time. Since organized societies have faced crisis, individuals have stepped up to protect their fellow human beings from harm. Only after our settlements became more complex and more densely populated did the more specific profession of the urban emergency manager emerge.

As a formal specialization, emergency management has evolved in fits and starts. And reviewing where we have been often provides important insight into where we are going. In the United States, this means assessing the civil



Urban Emergency Management. DOI: http://dx.doi.org/10.1016/B978-0-12-805307-2.00014-7

defense efforts of the 1950s and the 1960s, and the arrival of the Federal Emergency Management Agency (FEMA) in the late 1970s.

FEMA had responsibility and authority for a range of hazards, from severe weather to forest fires to nuclear fallout. Critically, the agency head reported directly to the Executive Branch of the federal government, and emphasis was placed on forecasting, preparing for, and mitigating future emergencies. During the 1980s, FEMA was largely tasked with Cold War functions such as continuity of government. FEMA moved into a unified headquarters and founded the multisector training site in Emmitsburg, Maryland, which would evolve into the highly respected Emergency Management Institute (EMI).

The crucial Stafford Act of 1988 essentially replaced the previous civil defense regulation and guidance. The act authorizes major emergency and disaster declarations, sets broad eligibility criteria, and specifies the types of assistance available from the federal government (Sylves, 2008, p. 60). In the 1990s, FEMA administrator James Lee Witt placed an unprecedented emphasis on mitigation as a key part of national readiness. Witt was the first true emergency-management professional to hold the role, and changes he instituted included forming three major phase-based "directorates" within the agency: mitigation, preparedness, and response/recovery.

The Department of Homeland Security (DHS) was officially created by the Homeland Security Act of 2002, under President George W. Bush, and former Pennsylvania governor Tom Ridge was named as its first secretary. The general organizational format of the new agency was adopted in November of that year, and by March 1 of 2003, more than 175,000 employees from 22 existing federal agencies were formally folded into DHS.

The tenure of emergency response professional Craig Fugate as the director of FEMA has been an unqualified success. Fugate is widely respected within the emergency-management practitioner community, and the agency's planning and response efforts have generally received high marks. The National Incident Management System (NIMS) model of incident command systems (ICS) continues to expand and evolve, and the system has been widely adopted across the private and nonprofit sectors as well. It presages the general framework that will guide the US model going forward.

## Current UN, European, and Asian Models

The United Nations (UN) does a wide variety of development work that may be considered as aligning with the emergency-management phases of preparedness or mitigation, such as improving critical transportation networks, or providing fresh water supplies to reduce the risk of water-borne disease. The Emergency Response Division (ERD), Disaster Reduction and Recovery

Programme (DRRP), and Bureau of Crisis Prevention and Recovery (BCPR) were created within the UN's larger Development Programme (UNDP) operation. Specialized teams designed for prompt field response to disasters were allocated fully 5% of the overall UNDP budget (Haddow, Bullock, & Coppola, 2008, p. 257).

The crucial Sendai Framework, adopted by the UN in 2015 and 2016, has the potential to make a major and lasting change to the entire field of disaster science. Named for the location of the Third UN World Conference on Disaster Risk Reduction—held in Sendai, Japan, in March of 2016—the accords set goals for a 15-year period. The language places great emphasis on utilizing science and technology to reduce the impacts of future catastrophic events across the globe (Dickinson, Aitsi-Selmi, Basabe, Wannous, & Murray, 2016, p. 108–109). Among the conference results was a Road Map to 2030, which focuses on four challenges to be addressed:

- How to leverage networks and platforms to advance research joining science, policy, and practice
- How disaster risk is understood and assessed, and how warning systems are designed
- What data, standards, and practices are needed to measure and report risk
- Identify what research and capacity gaps exist, and what difficulties must be overcome

The European Union launched its Civil Protection Mechanism in 2001, a cooperative agreement offering aid and assistance across national borders. It was soon followed by the Emergency Response Coordination Centre (ERCC), an emergency operations center with extensive monitoring and communications capabilities. The EU effort focuses on natural and transnational hazards. The organization was crucial in coordinating the response to European flooding in 2013 and 2014 and to the African Ebola outbreak in 2014. Although the young mechanism is beneficial for member nations and others that request assistance, Europe still lags behind the more unified federal efforts of the United States.

The Association of Southeast Asian Nations (ASEAN) has also implemented a number of mutual-aid compacts over recent decades, including major agreements in 1967 and 1976. Most recently, the Vientiane Action Programme set increasing goals for cooperation and interaction from 2004 to 2010. Agreements are specifically worded to preserve national sovereignty, and stress joint response to disasters via intensified regional and international cooperation.

It is worth noting that the earthquake and resulting tsunami that stuck the Indian Ocean late in 2004 remains the largest and most catastrophic incident in modern times. However the world response was unprecedented in both scale and in the number of nations taking part.

Yet, as will be described, this model of international cooperation and interaction is under assault.

## The Current State of the Art

Overall, the current intellectual and operational state of emergency management could be accurately described as very active. The threats and hazards facing the global community have in no way decreased, and the root causes of natural, manmade, and hybrid disaster show no signs of abating. Once again, emergency management and evolving ICS methodologies will play a key role in facing these challenges today and in the future.

This text has repeatedly cited the 300 emergency-management degree programs in place on the scholarly side of the profession, and this number is expected to experience continued growth. The merging and interaction between the operational and academic branches of a given field are unquestionably one of the best predictors of its health and vibrancy moving forward.

Professional emergency managers will have three key assets at their disposal in an increasingly complex world: personnel, technology, and research and education.

## The Way Forward

### *Personnel*

The role of an urban emergency manager is rapidly evolving. As discussed, the profession requires top performers to be both generalists and subject-matter experts. A strong basis in all-hazards management is the foundation, and expertise in specific interests within the field adds a highly desirable layer of depth.

There is literally a generational shift occurring within the profession. As emergency management and homeland security came of age around the turn of the century, particularly in the United States, the vast majority of frontline practitioners were transfers or retirees from related fields such as law enforcement or the fire services. This first generation of professionals is reaching their final retirement age and are, in effect, passing the baton to the next generation.

The new generation of emergency-management personnel will quite literally set the tone for the next two or three decades. This represents an incredible opportunity for the field to go through yet another rapid evolution as younger, career-minded, and tech-savvy professionals take the helm. Acting as a counter balance is the fact that federal, state, and local funding for emergency-management staffing hit a peak soon after the September 11 attacks and has been slowly descending since. This troubling trend will continue until the next major disaster occurs, once again drawing the fickle attention of politicians and the public.

For a truly successful transition, this new, leaner generation must learn the hard-earned lessons of their predecessors, and also bring a diversified and fresh perspective to the role. The future will be no less challenging than the past; in fact, in many ways it will be more so. Technology, education, and qualified personnel will all play an immense part in this truly transformational shift.

### *Technology*

Perhaps no single factor will have a greater impact on the future of emergency management than that of technology. This broad category includes the equipment, facilities, and systems that emergency managers utilize to perform their roles.

Young people, including many emergency-management students, perceive technology as a beneficial and benevolent force. They have grown up in the digital age, surrounded since their childhoods by the Internet, smartphones, and touchscreens. Multiple studies show this age group to hold very positive views about technology.

Older people, including most seasoned emergency managers who may have transferred to the field following other public-safety careers, have a much less rosy perception. In many instances, technology has been a disruptive force—not necessarily in a positive sense—requiring them to learn and relearn the same functions as new systems and software come into favor. Multiple studies show this age group to hold less positive views about technology.

As a mid-career professional, the author, somewhat predictably, falls somewhere in between. Technology has indeed been both a benefit and a burden since the advent of the digital age. And there is no indication that it will be purely a dream or purely a nightmare going forward. Rather, it is likely to both benefit and threaten emergency managers in the future.

Two technology categories likely to have a particularly large impact on emergency management in the coming years are digital communications and so-called cloud storage.

Communication is rapidly becoming both easier and more complicated due to the digital revolution. Cellular smartphones, Wi-Fi networks, and reliable digital radios have already made an immense positive impact on the phases of response and recovery and will continue to do so. Mass messaging and alerting services have become easier and less expensive for municipalities and organizations to employ.

Similarly, "cloud" storage has revolutionized the way data are saved and retrieved. Rather than storing information on local drives, cloud computing allows the efficient storage of large quantities of data on massive "server farms" for easy retrieval via the Internet. For emergency managers, this means emergency plans, floorplans, video footage, and contact lists are all much more accessible than in the past via a quick download.

Other technologies such as unmanned vehicles (drones) and high-speed data analysis (big data) will also have outsized impacts on current capabilities. All of this convenience and flexibility comes with a price, however, as will be assessed later in the chapter.

### *Research and Education*

The third crucial asset at the disposal of future emergency managers is that of research and education. As an emerging field of study, emergency management is undergoing explosive growth. As of 2016, FEMA recognizes 310 separate degree programs at nearly 100 different colleges and universities. And that figure is merely within the United States.

As a result, emergency management is emerging as a viable and active field of research, attracting top academics and spawning multiple professional, peer-reviewed journals including the Journal of Emergency Management, the International Journal of Emergency Management, and the Journal of Homeland Security and Emergency Management. Such vibrant scholarship inevitably transmits from academia to operations, as it has in countless other fields.

Professional organizations such as the International Association of Emergency Managers (IAEM) also provide valuable opportunities for learning and networking as well as for professional certifications in the field. Taken together, these developments represent an active and thriving academic component within the field. High-achieving emergency managers will avail themselves of this flurry of scholarship and operational knowledge, and hopefully drive the crucial cross pollenization between the two.

## FUTURE TRENDS

Despite warnings to the contrary at the beginning of the chapter, any holistic text on urban emergency management would be remiss without making some forecasts on the future of the profession. A few key components are clear.

## Technological Changes

As one of the three pillars of growth in the profession, technology is a key driver of change. The aforementioned digital communications and cloud computing will see massive shifts in the coming years, but they are not alone. Technology will remain a double-edged sword within emergency management's foreseeable future.

Vulnerability to cyber threat will continue to accelerate and morph. As the so-called Internet of Things (IoT) makes more and more devices, both portable and static, connected to the web, it also exponentially magnifies the exposure of not just those devices but entire networks to malicious actors such as a pair of massive distributed denial of service (DDoS) attacks in late 2016. The current digital-security protocols for such devices are horrifically poor, and the importance of this status going forward cannot be overstated.

In addition to the expanding range and number of tech-driven objects connected via the Internet, we may anticipate a growing presence of malicious users. Various totalitarian regimes have engaged in organized programs of political harassment and misinformation as well as used state resources to accommodate the large-scale theft of intellectual property. Given the minimal risk of meaningful repercussions, emergency-management professionals should anticipate similarly disruptive actions to be taken in the near future by individuals or small groups with ties to organized crime and terrorism as the technology and know-how continues to spread. Emergency managers must advocate for critical systems to possess the latest and greatest capabilities in resiliency and information security.

Cloud storage has also done little to address data-security concerns. Emergency managers should comport themselves with the understanding that anything they place online or in cloud storage may someday be viewed by the general public. It is urgent that vital documents and plans be backed up in multiple locations to avoid deletion, whether accidental or intentional. There can be little doubt that emergency-management data—particularly with an overlapping homeland security component—will be targeted by malicious actors, including foreign governments and intelligence services.

The presence of mobile or cellular phones has also been a comparatively recent development, greatly aiding both victim and responder communication. Cellular location finding will continue to improve and expand. Next-generation public-safety answering points (PSAPs) will be equipped with far-superior location technology, enabling dispatchers and responders to locate victims more quickly and efficiently. PSAP facilities will also be able to receive additional information via text and digital photos. Emergency managers in the field must be aware that smartphones also mean they are on view constantly. Be cautious in the knowledge that video showing a bout of laughter or a minor disagreement among responders at a disaster scene could be taken completely out of context and go viral in a matter of minutes.

Unmanned vehicles or "drones" will also have an enormous impact on emergency management in the coming years—both positive and negative. On the positive side, aerial drones will come into more common usage in operations such as missing-person searches, wildland fire reconnaissance, and structural inspections. Ground-based drones will become more prominent in search and rescue, firefighting, and counterterrorism operations. These uses will benefit both manpower requirements and safety by reducing the number of human beings placed in harm's way to accomplish mission objectives. On the negative side, there have already been multiple cases of manned firefighting aircraft being placed in danger by uneducated or incompetent civilian drone operators. As the devices become ever more ubiquitous, this occurrence too will become increasingly common at disaster scenes, including within the urban environment.

Virtual and augmented reality (VR and AR) are other fascinating technologies gaining expanded use. After years of promise, VR systems are finally coming online in affordable and useful applications. Training in particular will be revolutionized by much more accurate simulations of real-world situations. The very near future will witness new and seasoned firefighters training in virtual infernos, and police officers being placed in multiple simulated life-threatening scenarios before ever reaching the street. Emergency managers will likely benefit too, as incident-management training is poised to take a digital leap forward.

AR is equally promising as it moves into field usage by responders and planners. Imagine a firefighter having a detailed digital heads-up display inside his or her helmet, feeding real-time information such as ambient air temperature and remaining oxygen quantities. Or envision an emergency manager standing at the forward command post surrounded by a three-dimensional model of the affected facility's most recent floorplan.

Ultimately, as it has been throughout history, emerging technologies will be neither a purely positive nor negative force. Rather, they will serve to increase the capabilities of emergency managers and contribute to new and evolving risks as well.

## Political and Social Issues

Emergency managers must, of course, operate within the broader context of societal and political realities. Predictions in this realm are particularly fraught, given the instability across Western democracies.

Government gridlock has been a grave concern in many nations, particularly in the United States, where for more than a decade a dysfunctional Congress could rarely pass an annual budget without extreme drama—often including short-term shutdowns of the federal government. The lack of effective planning and funding for the Zika virus outbreak which began in Puerto Rico and Florida in 2016 is a prime example of a known threat being ignored by politicians who are willing to put public health at direct risk for partisan reasons.

Executive emergency-management leadership must also be qualified and competent professionals rather than purely political appointees. "You're doing a heck of a job, Brownie" scenarios must not and cannot be repeated. The confirmation of Brock Long, an emergency management professional with both state- and private-sector leadership experience, as FEMA administrator in the summer of 2017 indicated a promising development.

The role of US president is in many ways that of "emergency manager in chief." One a smaller scale, this title also applies to state governors and to big-city mayors. However, even highly skilled frontline emergency managers find it hard to operate in such a politicized funding environment. As Sylves astutely notes, "part of the politics of disasters comes in the form of both competition and mutual accommodation of the partisan behaviors of elected executives" (2008, p. 215).

And much will certainly depend on the burgeoning trend of nationalism and isolationism seen across multiple continents. In just a few short years, nations as varied as Hungary and the Philippines have elected extreme right-wing authoritarian leaders and voters in Britain chose to abandon ties to mainland Europe. Other developed nations such as Austria, France, the Netherlands, Turkey, and Poland are simultaneously witnessing virulently nationalistic movements. The unfolding presidency of a similarly aligned figure in the United States now represents nothing less than a complete seismic shift in the vision and policies of the world's sole current superpower.

It is far too early to tell what the long-term impact of such populist waves will be upon the field of emergency management as a whole, and the initial indicators are decidedly mixed. Extrajudicial killings in the Philippines have skyrocketed, particularly in urban areas. International agreements and cooperation in England and Hungary have virtually ground to a halt as core domestic and foreign policies are frozen or scrapped outright. By contrast, potentially beneficial infrastructure renewal programs have been discussed and promised in various nations, but not yet broadly implemented. Ultimately, the international leadership of the coalition of Jeffersonian/Hamiltonian democracies led by the United States is in question in ways it has not been in over a century, which has the very real potential to lead to spike in a range of serious global issues such as cybercrime, pollution, and international terrorism. Politically, this is a time of enormous unrest.

From a purely apolitical perspective, this matters deeply to emergency managers because such governments and administrations are elected in part due to promises to slash international cooperation and drastically reduce or eliminate spending within the public sector. Both actions have a direct and chilling effect on coordinated regional emergency-management operations. In the near-term, this will make efforts such as the 100 Resilient Cities program all the more important, as a lack of national and federal cooperation will make peer-to-peer and professional interactions even more vital. Many functions are likely to shift to nonprofit or volunteer organizations out of utter necessity.

Running in complete contrast to these developments is the issue of public expectations. In North America and Western Europe, citizens have an extremely high—some may say unrealistic—expectation that governmental agencies will prevent or quickly correct every possible emergency. Any violent act of terrorism or extremism is viewed as a preventable failure. Homes and businesses are built on oceanfront lots primed for destruction by storm surges, or on steep, wooded hillsides in the direct path of wildfires. And when the inevitable occurs, emergency managers are expected to ride to the rescue with hordes of resources, buckets of money, and absolutely no delay.

When FEMA, for example, stated that citizens needed to be prepared to be self-sufficient for 72 hours following a disaster, it got considerable pushback from the public, from politicians, and from the media. The flip side of this coin is that the public's attention span can be quite limited. If they have not experienced a recent disaster in close geographical proximity, residents (and taxpayers or shareholders) can be quite skeptical of supporting and funding emergency management programs. Effective emergency managers must therefore also strive to set realistic expectations among their constituents. Stability has always been a luxury in the world of emergency management, perhaps now more so than ever. Ultimately, it is always better to underpromise and overdeliver.

## Systemic Changes

Several important structural trends in emergency management may be expected to continue or possibly accelerate. Looking ahead, these factors have the potential to rise to the level of true system change in the field.

The use of contractors by both government agencies and large organizations is likely to continue in the coming years. For both fiscal and liability reasons, contract specialists make sense for many operations. These include, but are in no way limited to, logistics, call takers, communications, recovery logistics, and hazardous materials handling. Consultants will be a continuing presence in both the emergency-management and homeland security fields, as long as federal governments and large corporate entities continue to dedicate funding to such project management.

The use of military reserve or National Guard units will also continue and possibly expand. The level and types of training provided to these units will also likely change and expand in the United States, as they will be deployed domestically for large-scale severe weather events or acts of terrorism. The post-2001 advent of the new Northern Command indicates that active duty military units may also become more involved in disaster response as state and local leaders seek additional assistance from the federal level. In the United States, this may even mean some adjustment or cautious rewrite of the current posse comitatus legislation.

Terrorists will continue to attack soft targets for the foreseeable future. Even if the Islamic State in Syria and the Levant (ISIL) continues to lose ground and territory on the conventional battlefield, or is completely defeated as a standing army, it will also continue to lash out at Western cities. Its current, relative capabilities do not presently allow an attack on the scale of September 11, but there will be a steady hum of violence like that seen in Paris, Nice, Brussels, and Berlin. An active sociopolitical debate is taking place across Europe regarding the balance between security and freedom of movement (Birnbaum, 2016). Indeed, in Europe the primary threat will continue to come from Islamic extremists for the many reasons cited in Chapter 8, Emergency Management and Homeland Security in Urban Areas. The likelihood for a similar, or even more extreme, group to fill the vacuum caused by any decrease in ISIL capabilities is very real.

Overall, terrorism and extremism will remain a real and constant risk to emergency managers and to public safety writ large. In the United States, the near-term threat will be from a combination of Islamic extremists and far-right antiestablishment extremists, both of which have easy access to explosives recipes on the Internet as well as to high-powered firearms. Unfortunately, homegrown violent extremism is likely here to stay.

The blending and blurring of emergency-management and homeland-security functions will also continue. As previously addressed, the overlap of hazard types means that many responses look alike—a chemical spill versus a chemical attack, for example. As the public sector in particular is squeezed to do more work with fewer individuals, alignment of such resources is inevitable. On the positive side, this will provide an opportunity for cross training and add tangible value to prepared and capable emergency departments.

On a related front, collaboration across sectors will become more vital in the near future. No government, corporation, or nonprofit organization has the resources required to plan for, respond to, and recover from every type of potential disaster. As a result, successful operations will make and strengthen cooperative agreements across sectors, jurisdictions, and even among competitors. For example, when a regional bank in the Great Lakes suffered a major fire during renovation construction in 2004, more than one competitor offered temporary office space through prearranged business-continuity agreements. Expensive or costly items, particularly highly specialized units or equipment, will be shared across sectors within urban areas.

It will also be interesting to see if a best practice emerges across jurisdictions and sectors in terms of emergency managers responding to the field. (See Chapter 12, Response Phase, for more detail.) At present, the professional community remains divided on the forward deployment of emergency managers and for what types of incidents. Much will depend on determining the most effective role for such valuable professionals, and this element of the debate could thus continue for years or even decades.

Pandemics will become an increasing and ever more unpredictable threat. This is due not only to the changing global weather patterns (addressed shortly) but also to a marked change in the ability of infectious disease to resist current antibiotic regimens. Rapidly emerging drug-resistant "superbugs" are far outpacing the pace of pharmaceutical development. This lag in available countermeasures results in an increased risk of a virus or bacteria spreading rapidly and disastrously through the human population. As Western governments focus much of their attention and funding on terrorism and other human-caused threats, it is worth recalling that the influenza outbreak of 1918 killed far more people than the world war that preceded it.

Taken as a whole, these structural trends will alter the very paradigm of urban emergency management in the years to come.

## A True Game Changer

The greatest shift facing every urban emergency manager on Earth is climate change. In the near-term, much of the planet will get warmer, some parts cooler. Some regions will see increased precipitation, many will see a steep drop in rainfall. What is very clear is that the long-term weather patterns for precious few urban areas will remain stable or unchanged.

Scientific models vary slightly, yet emergency managers must be prepared for an average sea level rise of between .5 and 1.25 meters (roughly 1.5–4 feet) in the world's oceans by 2100. This amount will largely depend upon the additional quantity of carbon that is placed into the air, and the speed at which the polar ice caps melt and recede. This increase bodes especially poorly for large cities built immediately adjacent to the world's oceans.

Public-safety officials from Miami to Naples, and from Caracas to Hong Kong, must put mitigation plans in place prior to potentially catastrophic-level impacts truly taking effect.

Additional levees, dams, pumps and other engineering projects will indeed blunt the full force of the rising waters. But they will not solve the problem entirely, and some populations will simply have to be relocated further inland. If this is not accomplished in a structured and well planned manner, it will take place regardless, and likely lead to widespread disruption, civil unrest, and commodity shortages. Governments in comparatively low-income, low-lying nations—such as Bangladesh—can anticipate nothing less than a complete societal disruption.

Droughts and wildland fires will also become more frequent, larger, and more widespread. Both Australia and North America have already seen this pattern emerge. Effective water and land management policies will be absolutely key to minimizing the impact of significant precipitation changes.

Finally, disease will also adapt and mutate as the planet warms, magnifying overall risk. Outbreaks of truly horrific communicable diseases such as Ebola become a very real threat. Mosquito-borne illnesses such as Zika and chikungunya will continue to expand their range and to impact ever larger human populations. The possibility of a pandemic only increases if uncontrolled mass migrations due to flooding, drought, or armed conflict take place.

If this future sounds especially gloomy, consider that mankind has made it through civil strife, world wars, and epidemics before. Qualified and capable emergency managers will be largely responsible for seeing to it that we do so again.

## CONCLUSIONS AND SUMMARY

The future is never a certainty, but multiple patterns and trends do allow some insight into the coming years. Technology, personnel, and scholarship and education will all be crucial factors in coming changes within the field. Corporations, nonprofit organizations, and governments will all have significant influence over the course of future events.

Politicians and social trends will continue to directly and indirectly impact emergency management. Contractors and military units will also have an increasing presence. And a changing climate is the massive wave of change that will impact not just emergency management but also the basic governance of cities the world over.

When all is said and done, the profession of emergency management remains a dynamic and exciting one for students, scholars, and practitioners. We can most ideally serve our fellow human beings by studying past disasters and by applying best practices and novel approaches to the ones that will doubtlessly occur in our lifetimes. For at least one thing is certain about the future: emergencies and crises will continue to occur, and people will be looking for the types of leaders capable of compressing the chaos.

## References

Birnbaum, M. (2016). Europe may face grim future with terrorism as a fact of life. *Washington Post*. Retrieved from: <www.washingtonpost.com>.

Dickinson, C., Aitsi-Selmi, A., Basabe, P., Wannous, C., & Murray, V. (2016). Global community of disaster risk reduction scientists and decision makers endorse a science and technology partnership to support the implementation of the Sendai Framework for disaster risk reduction 2015–2030. *International Journal of Disaster Risk Science, 7*, 108–109.

Haddow, G., Bullock, J., & Coppola, D. (2008). *Introduction to emergency management* (Third edition). Burlington, MA: Elsevier.

Sylves, R. (2008). *Disaster policy & politics: Emergency management and homeland security*. Washington, DC: CQ Press.

## Further Reading

Birkland, T. A. (2006). *Lessons of disaster: Policy change after catastrophic events*. Washington, DC: Georgetown University Press.

Hogue, H., & Bea, K. (2006). *Federal emergency management and homeland security organization: historical developments and legislative options*. Washington, D.C.: Congressional Research Service.

Roberts, P. (2006). FEMA after Katrina. *Journal of Policy Review, 137*, 15–33.

U.S. Department of Homeland Security (2011). *National preparedness system*. Washington, D.C.: Federal Emergency Management Agency.

# Index

*Note*: Page numbers followed by "*f*" and "*t*" refer to figures and tables, respectively.

## A



## B



## C

## O



## P

9780128053072